God's Tactical Unit
Spiritual and Temporal Warrior Parallels

God's tactical unit drawn together by the Holy Spirit's power;
Warriors gifted to witness throughout the earth to the last hour.
(Paraphrase of Acts 1:8)

David A. Darlington (M.Div.)

GOD'S TACTICAL UNIT
Copyright © 2013 by David Darlington

All rights reserved. Neither this publication nor any part of this publication may be reproduced or transmitted in any form or by any means, electronic or mechanical, including photocopying, recording or any information storage and retrieval system, without permission in writing from the author.

Unless otherwise marked, all scripture quotations are taken from THE HOLY BIBLE, NEW INTERNATIONAL VERSION®, NIV® Copyright © 1973, 1978, 1984, 2011 by Biblica, Inc.™ Used by permission. All rights reserved worldwide.

Printed in Canada

ISBN: 978-1-77069-777-5

Word Alive Press
131 Cordite Road, Winnipeg, MB R3W 1S1
www.wordalivepress.ca

Library and Archives Canada Cataloguing in Publication

Darlington, David, 1946-
    God's tactical unit : spiritual and temporal warrior parallels / David Darlington.
ISBN 978-1-77069-777-5

    1. Canada. Canadian Armed Forces--Religious life. 2. Soldiers--Religious life--Canada. 3. Military chaplains--Canada. I. Title.
UH25.C2D37 2012   355.3'470971   C2012-907598-1

DEDICATED TO

My parents
Josiah and Edna
May the legacy of your faith, hope, and love
live on through this book.

"Not to us, O Lord, not to us but to your name be the glory, because of your love and faithfulness." (Psalm 115:1)

ACKNOWLEDGEMENTS

I want to extend a heartfelt "Thank you" to Janice Henderson for innumerable hours of editing. Her suggestions have added quality to the book and her encouragement has been an immense inspiration.

Special recognition and thanks is also bestowed on Dianne Collier for her friendship, references, and advice on military life and the publishing process.

In addition, I acknowledge my indebtedness to the staff at Word Alive Press for addressing an endless stream of questions and polishing the final edition of this book.

Finally, my deepest gratitude to:
Canadian Forces personnel, chaplains, veterans and their families

Christian congregations in proximity to CFB Borden and CFB Petawawa

My family and friends for their continued interest, encouragement, and love as this book has unfolded.

You have all impacted my life in ways none of us may ever full realize.

TABLE OF CONTENTS

ONE Priority: Troop Training	1
TWO Mapping Past Strategies	19
THREE Boot Camp	39
FOUR War Declared!	65
FIVE Posted—Again!	77
SIX Commence Firing!	89
SEVEN For Better or Worse	103
EIGHT Guarding Our Own	129
NINE Returning Home	147
TEN Chaplain's Corner	157
ELEVEN Cease-fire!	177
Epilogue	187
Appendices	191
Bibliography	201

Priority: Troop Training
ONE

THE BOOK YOU HOLD IN YOUR HAND WAS DEVISED PRIMARILY AS A LIFE-changing weapon. It is a primer for life as *"a good soldier of Christ Jesus"* (2 Timothy 2:3) designed to enrich the spirit of individuals in God's tactical unit. God's people are spiritual warriors trained and gifted to dispatch the Good News of His love to their community and beyond. For some, the book may provide nothing more than a review; for others, a challenge to a deeper commitment in their Christian walk; and to still others, a revelation of possibly a new life in Christ Jesus.

The book's second objective is to raise awareness within Christian civilian communities of the need for Canadians, especially Christians, to enhance support for our troops. These targets are attacked by aiming at existing parallels between two methods of warfare identified as spiritual and temporal.[1] The strategic plan calls for assaults launched from two fronts. One thrust will engage readers' minds with paraphrased scriptural and non-scriptural poetic readings. This will allow readers to trigger the creative side of their brain while activating the rational side through the main thesis. Concurrently, a second invasion commences awakening the readers' spirit and soul to some identifiable parallels between biblical life and military life. The strategy engages members of God's tactical unit from the Old and New Testaments who exhibit certain lifestyle similarities to military families. Additionally, readers will be introduced to some of the

---

[1] Joyce M. Hawkins, compiled by. *The Oxford Paperback Dictionary: Third Edition.* Oxford: Oxford University Press, 1988. 842. *temporal (adj.)* 1. secular, of worldly affairs as opposed to spiritual.

unique "Militarian" dialect. One key for understanding a culture is to comprehend their language. In military life, every conversation includes its own idioms employing an extensive use of Greenwich Time and a variety of acronyms. A number of the acronyms are utilized throughout this book following their initial presentation.

It is time to put a rifle to my shoulder and fire two opening shots in question form. First, are you a Christian? Second, if not, have you ever wondered what the Christian life might be like? There are only three options when answering these questions. One can reply "Yes" and "Yes, I know what it means to live as a Christian." A second response could be "No" and "No, I don't care to know," in which case you may not be interested in reading further; but I hope you will read a couple of chapters and maybe beyond. The final possible answer is "No" and "Yes, I'm curious about what makes Christians tick." This book is offered to assist individuals who have declared war on their soul's enemy and those who would consider such action. If you have decided to place Jesus Christ as Lord over your life and accepted Him as your Saviour who removes sin, you may already be aware of your battle with Satan. You are already engaged in spiritual warfare. If the God of the universe is of no interest to you, then Satan is quite content not to cause trouble because you are in his camp. If you wonder what Christian life involves, you can start to develop an idea of the nature of its lifestyle by reading this book.

I also need to be honest and up-front with my readers. When one decides to accept Christ as their Lord and Saviour, they automatically place themselves in God's tactical unit; that is, His skillfully planned group of individuals called to battle in spiritual realms. In addition, even the thought of taking a step of faith in Christ's direction places you at odds with the forces of evil. I want my readers to fully understand there are only two operational forces in the spiritual and temporal realms: good and evil.

I contend temporal engagements are actually spiritual battles played out in the worldly realm. This means good and evil are also the only forces operational in the temporal or non-spiritual world in which we live. In support of this idea, Cole credits Origen (A.D. 185–254) as claiming: "Christians supplied an irreplaceable aid to the emperor.

By overcoming in prayer the very demons that cause wars, Christians actually help more than soldiers."[2] He contends they may not have gone on campaign with the emperor but did go to battle for him in prayer. As well, Cole notes, "What Augustine (A.D. 354–430) bequeathed to the Christian doctrine of just war was a clear understanding that human sin makes necessary the use of force."[3] In other words, these two fathers of the early church connect demons and human sin with warfare. To state it another way, the forces of evil call forth legions of demons against all who enlist or consider signing up in God's tactical unit. It was the Apostle Paul who urged Timothy to: *"Endure hardship with us like a good soldier of Christ Jesus. No one serving as a soldier gets involved in civilian affairs—he wants to please his commanding officer"* (2 Timothy 2:3–4). Trouble and hardship are part of life serving in God's tactical unit under Jesus Christ, the Commanding Officer (CO). They are exposed here because nothing is gained by hiding these two strategies employed by the enemy of our soul; I call these his improvised explosive devices (IEDs). Paul was privy to two mysteries that many since have discovered. First, the enemy was defeated by Christ's sacrifice on the Cross; yet he persists in his menacing way *"like a roaring lion looking for someone to devour"* (1 Peter 5:8). Second, Paul claimed two promises of God: *"I will never leave you nor forsake you"* (Joshua 1:5) and *"My grace is sufficient for you, for my power is made perfect in weakness"* (2 Corinthians 12:9). These were fulfilled in Paul's life and are available to Christians today. God is not a spectator in the bleachers cheering His army on; He is a strategic member of His tactical unit found in the midst of every battle—either spiritual or temporal. In the temporal battle to overtake Jericho, the Ark of the Covenant and the priests were a personification of God. He placed Himself between the armed guard and rear guard. In modern warfare, chaplains are God's ambassadors who carry His Word to the front line, while His *"chosen people, a royal priesthood"* (1 Peter 2:9) share His Word on the home front. God is central in warfare whether it is viewed as a spiritual or temporal conflict.

---

[2] Darrell Cole. *When God Says War Is Right: The Christian's Perspective on When and How to Fight.* Colorado Springs, CO: WaterBrook Press, 2002. 11
[3] Cole. 23

As I begin this troop training, four topics of 'classified' information require my attention in a civilian environment; however, this would not be so in military boot camp, as will be revealed later. Civilian life, unlike military life, is full of demands for information and questions requiring answers. I suspect readers are presently curious about my life, the need for this book, what biblical and historic roots it is based upon, and what differences exist between life in the military and civilian world. We will deal with the first two items of 'classified' information in this chapter and close with an overview of military life. The second chapter reveals biblical and historic roots, which cover a broad span of time. The balance of the book turns its focus on military lifestyle issues. 'Starting Point Unknown' is one of numerous poetic creations you will discover as you read the book. The poem is followed by a brief testimony of my life and ministry including military and other related experience.

STARTING POINT UNKNOWN

What was the starting point for your life or mine?
The answer we can find in Psalm one thirty nine.
'For my inmost being You certainly did create,
In my mother's womb with Your skill so great.
Fearfully and wonderfully fashioned was I.
Knowing this full well! You, I will glorify.
My frame, not hidden when at a secret site made;
My unformed body did not Your sight evade.'[4]

Now before our life began, God had His own plan spun.
He tells of His foreknowledge in Jeremiah chapter one.
The word of the Lord came to me and stated:
'I knew you before in the womb you I created.
Yes, before you were born I did set you apart;
Appointing you to the nations with a prophet's heart.'[5]

---

[4] Paraphrase of Psalm 139:13–16
[5] Paraphrase of Jeremiah 1:4–5

Because God's love for us was so great,
He gave us free will to decide our own fate.
One criminal beside Christ waited to the eleventh hour
To cry, 'Remember me in Your kingdom and power.'
For his show of faith, Jesus replied: 'Today you will be,
I'm telling you the truth, in eternal Paradise with me.'[6]

Full of the Holy Spirit and faith, Stephen was the man;
But Saul, later Paul, of his stoning approved the plan.
Then on the way to Damascus, the Lord called to Paul,
His chosen instrument to reach Gentiles, one and all.[7]
Jeremiah, the criminal and Paul—at different life stages
Turned from sin and received forgiveness through all ages.
A free will choice, Christ to follow with all their heart.
This was the point where their real life found its start.

An unknown starting point but from God not hidden.
He planned life, set us apart and now we are bidden.
The starting point for all is to walk in Christ's way,
To do as He did and pray for right words to say.
When to begin is our free will choice to make,
But to ignore Him would be a sinful mistake.

The Author

In a number of ways, the above poem and the next one reflect the first forty years or so of my being. Childhood stories shared by my parents point clearly to God having a plan for my life; it could be argued this plan was even spun before my conception. In teen years and beyond, I followed my own path—one of doing things *my* way. Those were not years of moral corruption but a season of rebellion against God, which included random church attendance, marriage, starting a family, and beginning *my* dream to operate and own a dairy farm. Those years produced financial struggles, near bankruptcy, family breakdown and

---

[6] Paraphrase of Luke 23:39–43
[7] Read Acts 6:5, 8:1, 9:1–19

eventually divorce. All of these IEDs resulted in a total surrender to the God of my parents whom I knew but choose to try to serve *my* way. Once I decided to totally commit *my* life to Christ, the Holy Spirit (Commander-in-Chief) directed and guided my life as a roaming preacher. My new life mission involved years of formal theological study as a mature student, various ministry roles, and ample opportunities to interact with members of our Canadian Forces (CF). Over the past twenty-five years, the Lord has opened doors allowing me to provide interdenominational ministry throughout Ontario. This has included serving in two regions near Canadian Forces Bases (CFBs). I continue to serve in a number of church settings around the Alliston/Barrie/Collingwood triangle with CFB Borden reasonably central. This ministry has been in addition to what I classify as a three-and-a-half-year deployment to the Ottawa Valley, with CFB Petawawa being at the Valley's heart. Invitations to various church and para-church ministries have made the CFBs within each geographic region more accessible. By serving ministries with different denominational loyalties, I have dialogued with a comprehensive cross-section of active and retired military personnel from these CFBs.

Conversations with chaplains posted at each CFB have furthered my knowledge of military life, accompanied by in-depth discussions with members of Military Christian Fellowship of Canada (MCF) from these bases. I also had opportunities to attend MCF Annual General Meetings in Ottawa, and to speak with the designated chaplain general of the Canadian Forces and with a vice-president for the Association of Military Christian Fellowships (AMCF), MCF's international affiliate. For a civilian, my diverse missionary-style ministry and extensive military interaction has accomplished two tasks. First, it raised my personal awareness of the need for Christian civilian gatherings to support our troops. Second, it confirmed my belief in a possible partnership harmony between Christian civilian leaders and the chain of command from the chaplain general and MCF executive on down. This is conditional, of course, on mutual respect for military ministry's uniqueness. Furthermore, the openness I discovered reflects conditions at the time of this investigative process and is subject to change. Part of the outcome of my life's mission is this book, which I believe the Holy Spirit commissioned. Disobedience to His command is never an option! This is the command given: Using your experience, file a report on the need for military ministry in Canada. Although He provided guidance and direction, I am responsible for errors or misinformation provided in the book. Admittedly, it is a fallible report and not necessarily the only approach for enriching the spirit of individuals in God's tactical unit, or for raising awareness of the need to support our troops.

When the Israelites reached the border of the Promised Land, Moses sent twelve scouts into Canaan who noted a powerful people well prepared for an attack. The scouts could not agree on a unified plan of action; ultimately, Caleb and Joshua disagreed with the other ten. But Caleb did report to Moses: *"We should go up and take possession of the land, for we can certainly do it"* (Numbers 13:30). This book may not receive majority support; in fact, it might be rejected as was the report by Caleb and Joshua. However, I evaluated the territory as one who journeyed into the realm of our military, a powerful people group, and now file this report. According to my assessment, the Christian civilian community can certainly, in cooperation with the Christian military community, go up and take possession of these lost souls (to paraphrase Caleb). As we know, Israel's failure to move forward resulted in forty years of wilderness wandering. I challenge my readers with this question: can the Christian church in Canada afford that type of delay today? I hope this brief account of my first sixty-six years of life has addressed some of the questions regarding my life, ministry, and interaction with our military.

As a wrap-up to this sub-section, I would remind readers that this is a subjective paralleling of two lifestyles to emphasize the need for military ministry. Additionally, it is an opportune point to parallel my life with some aspects of military life to be explored later. I was born in 1946 and come to this study with life experience, but not in a military sense. My father died from a brain tumor when I was seven years of age and I was raised by a godly single mother. As a baby boomer, I have not encountered war or even a genuine threat of war. I have made a number of moves, but only one required a major relocation and it still kept me in Ontario. There has been one deployment, as I describe it, to the Ottawa Valley for three and a half years of ministry. I struggled financially on the farm; went through the valley of separation and divorce; suffered with depression, suicidal thoughts, and reintegration into society because of the divorce; and assisted in raising two children as much as living apart and time allowed. Each of the above life experiences is common in military life. In my life, each of these encounters arose while I was associated with dairy farming, another unique culture in Canadian society. Agriculture, military, and sports are three areas used for illustrative purposes by Jesus and Paul. Are these three connected in some way, or was it because of people's familiarity with the activities they were used? When studied, all appear to share some common attributes. I claim military illiteracy, but the above disclosures point to somewhat-related experiences allowing me to interpret and make limited sense of what is critical for grasping the characteristics of military life. If you still discover that not all your questions are answered concerning my life, ministry, and military involvement, additional information will surface throughout the balance of this chapter and book.

THE HUMAN NEED

A wicked life freely chosen that shouts out: I'll do it *my* way!
But God's plan is very plain to see; yet we would rather stray.
His eternal power and divine nature, easily seen in what is made;
Heavens and new birth surround, so no excuse have we to bade.
But it is on the downward path; our society most often does take.
With foolish hearts we fail to praise His name, to our mistake.
In so doing, we like fools exchange the immortal God of glory,
For human comfort and pleasure, we ignore the entire Bible story.
Giving way to the pleasure of sex, and our heart's sinful desire,
Forsaking the truth of God's Word; we sink deep into the mire.
Worshipping and serving things created is the desire of our heart;
Ignoring the Creator and all He called *"very good"*[8] from the start.
Downward trends continue, as women give up a God-given desire;[9]

---

[8] Genesis 1:31
[9] Paraphrase of Genesis 3:16

To satisfy a lust on the prowl they go, for a woman to quell the fire.
With thinking gone faulty, the lie women accepted, so men do too;
In time gain corruption's penalty, as indecent acts with others they do.
With the depraved mind developing, doing what ought not be done.
We love greed, envy and strife; becoming a wicked daughter or son.
As gossip, slander and boasting most often spews off our tongue;
Faithless, heartless, ruthless and rebellion become our songs sung.
For depravity great, death is the reward deserved at an early date,
And our approval of others in sinful practice will just seal our fate.[10]

Knowing the reality to come, God foresaw our human need. How nice!
Thus Christ took on our curse substitutionally, His love paid our price.
The key is faith in Jesus Christ for all who would be saved. So hark!
For all have sinned to break God's heart by falling short of His mark.[11]

Not from yourself or by your works, thus no boasting you can do;
There is a gift by God's grace; it comes simply by faith to save you.
For we are God's workmanship, created in Christ Jesus to complete
Good works, which God prepared in advance for us to do; how neat![12]

The Need

As we move on with this introductory training session, the focus shifts to the need for this book. Early in the spring of 2009 while serving in the Ottawa Valley, I decided to poll church leaders in the region from Pembroke to Deep River. A summary of the questionnaire and survey is provided in Appendix A. Forty-seven surveys were circulated with a stamped self-addressed envelope directing them to my home address. I received thirty-two replies, providing a respectable return rate of sixty-eight percent. Each respondent had the opportunity to remain anonymous along with the option to omit statements or questions unrelated to their ministry, and some leaders exercised these privileges. I trust the region is representative of others in close proximity to a CFB

---

[10] Paraphrase of Romans 1:18–32
[11] Paraphrase of Romans 3:21–23
[12] Paraphrase of Ephesians 2:8–10

and hope it is typical of national interest in our troops and their families. Ten statements asking leaders to score their level of agreement with each one made up the primary component of this survey; the range applied being from one (disagree) to five (agree). I decided a three or higher on the scale represented agreement in principle with the statement. The statements were reduced to their basic point and agreement percentages calculated as recorded below. This resulted in the following:

> Summary of Agreement in Principle Statements
> 1. Churches viewing the military as a mission field: 100%
> 2. Churches need to be more supportive of our military: 100%
> 3. Churches need a plan to reach the military: 100%
> 4. Churches need a specialized military ministry: 83%
> 5. Churches would benefit from more information on the military: 100%
> 6. Churches interested in a needs and lifestyle seminar on the military: 85%
> 7. Churches interested in developing or expanding their military ministry: 87%
> 8. Churches believing support from military families attending their church would be beneficial for reaching other military families: 96%
> 9. Churches agreeing they may not be sensitized to military needs or may feel intimidated by the military: 82%
> 10. Churches seeing a community/base coordinator or liaison beneficial: 90%

When I conducted the survey, my thinking favoured a seminar but later developed into this book format. After analyzing the results, I found Christian local civilian community leaders seemed to be saying,

> We have a mission field at our back door that requires more support from us and our denominational affiliation, but we need a plan for reaching out to that community. It may not require creation of a specialized ministry, but we do need more

information on how to do what needs to be done. We have an interest in increasing ministrations to the military, and a seminar [or book] may help in the process; in addition, support from military families attending our church would be very beneficial. We may not have been sensitive enough in the past to our military or maybe we felt intimidated by them. Whatever shortcomings there may have been, we believe a liaison between the Christian local civilian community and CFB Petawawa could be of assistance in our pursuit of ministry to troops and their families.

In other surveyed sections, respondents exhibited interest through current projects and creative new initiatives suggested. The Christian local leadership unmistakably expressed a clear interest in extending their ministry to the military in their midst. These results parallel my impressions as a member of Pembroke and Area Clergy Association (PACA), the local ministerial. Please note, however, the survey was not restricted to PACA members but dispensed throughout the surrounding area.

There are two additional units within what can be classified as the Pembroke and Petawawa Christian community. These are the base chaplains and their programs, along with the MCF-Petawawa chapter (MCF-Pet). During my ministry, as stated earlier, I have held ongoing discussions with each of these and they are keen to work with Christian local gatherings. For example, MCF's representative at CFB Petawawa (MCF Rep-Pet) speaks to various church groups and other area organizations on a regular basis. The MCF Rep-Pet strives to increase civilian awareness of the complexity of military life. Just prior to the completion of my deployment, base chaplains introduced a new initiative. In an effort to expand their ministry to troops, the chaplaincy program inaugurated a mentoring program for civilians interested in military ministry. This type of action is supported from the top of these organizations down through their ranks. It even extends beyond the Canadian border, as AMCF sponsored a conference focused on military family life and ministry in a variety of areas at Estes Park, Colorado, in August of 2009. From reports at the MCF Annual General Meeting in the fall of 2009, this was an event well worth attending; regrettably, I was unable to do so. From my research, I have also discovered Campus Crusade for Christ International has a massive military ministry program in the USA. In summing up these scant accounts, they provide sufficient data to conclude there is a keen interest in military ministry by the key players, at least in Pembroke and surrounding area. Interest by the Pembroke/Petawawa churches, chaplains at CFB Petawawa, and MCF Rep-Pet indicate a need exists. Additionally, their initiatives demonstrate a willingness to cooperate to raise awareness of the need to support our troops. I pray this fire has or can be spread across our nation.

Prior to an overview of the military lifestyle, I offer readers an opportunity to reflect on my understanding of the Christian lifestyle.

THE CHRISTIAN LIFESTYLE

So what might my life as a Christian could it be?
Healthy, wealthy and worry free as some do see?
This prosperity gospel it certainly does abound;
But can its truth in Scripture be properly found?
Of Job, "he is blameless and upright" God has said.
Still many troubles did tumble down around his head.[13]
Tested by Satan, still in the end all turned out divine;
His spirit it was that God chose like gold to refine.
Or think of others that have been mentioned from the past.
What about Stephen, the martyr, whose life ended too fast?
Then there is Saul or Paul as did later become his name;
A "soldier of Christ Jesus,"[14] apostle and servant of fame.
But where was the health and wealth for this man so loyal?
For it was with lashings and beatings his life they did spoil.[15]
And yet he could claim: 'What was to my profit is but loss
Compared to the greatness of knowing Christ and His Cross.'[16]
Searching through the Word many more we could read about:

---

[13] Paraphrase of Job 1:8 & 2:3
[14] 2Timothy 2:3
[15] Paraphrase of 2 Corinthians 11:23–29
[16] Paraphrase of Philippians 3:7–11

As three in a furnace, one in a lion's den; still all walked out.[17]
Near life's end, John to Patmos as a prisoner they did confine;
Had a Revelation of end times and everything works out fine.
    As we walk through life waiting for the end to arrive,
    Attitude is the condition which most impacts our drive.

So what will my life be like? Consider the Lord you will follow:
Heaven's glory to nothing He became; a pill difficult to swallow.
Humbling him self—knowing his fate being death and pain;
Obedient to his Father, to death He did go and was slain.[18]
    Thus, should you expect a life filled with great ease,
    Or strive for one which will God fully please?
    As a child of God, for the latter you should hanker;
    For whatever befalls you He'll be your rear-flanker.
On Paul, God bestowed a promise: *"My grace is sufficient for you, for my power is made perfect in weakness."*[19] For us it is given, too!

The Lifestyle

Before wrapping up our training session, we turn to the differences between military and civilian life. Years of association with military personnel have resulted in learning on my part, especially in the area of lifestyle and its associated stresses. I will simply offer personal observations to provide a general overview of our troops and their families. Please remember you are hearing my voice sharing what I noticed and experienced, filtered through my personal grid of biases and emotions; this is not a scientific or sociological portrait.

    A major percentage of our troops are eighteen to thirty years of age, and many are married or partners in a common-law relationship. Of these, a large number have one or two pre-school children, and most live half the width of Canada or more away from their immediate families on one or both sides of the relationship. Over my years of association with military families, many of those I interacted with have been posted

---
[17] Read Daniel 3 & 6
[18] Paraphrase of Philippians 2:5–11
[19] 2 Corinthians 12:9

to other CFBs. This illustrates the difficulty in building a long-term friendship even in military circles, although technology does help. In short, moving or postings are a reality of the lifestyle. What does this imply for ministry? It signifies leaders will discover military ministry resembling a revolving door, as it is constantly in a state of motion. In this regard, it parallels youth ministry. The length of a military posting to one base has generally been no more than two or three years; fortunately, signs show this is starting to lengthen. Nonetheless, this gypsy lifestyle of military families not only impacts ministry, but second-income opportunities for a spouse as well. Local civilian employers realize a military spouse will be a short-term employee; thus, they tend to be reluctant to hire them. In addition, many young civilian and military couples lack budgeting skills, leading to poor money management and communication problems in their relationship. This is frequently compounded by peer pressure. When one adds all of these to the difficulty of securing a second income, they produce a climate for relationship failure. Combining these issues often generates arguments, fights, and occasionally physical and/or substance abuse. As one can imagine, substance abuse generates added financial burdens and continues the spiral on a downward cycle. Those who protect our freedom are just as prone to this cycle as those who enjoy that freedom. I have discovered two other issues. First, once a contract has been signed, the military in effective owns the member. Resignation is not an option. However, the military can issue a dishonorable discharge, causing an arduous stigma for future civilian employment. In other words, one cannot quit the military but can be dishonorably fired. But let me make it clear, a dishonorable discharge is never issued without justifiable reason(s). Second, divorce rates among military couples exceed the national average. This lifestyle places a great deal of stress on marriage and the family, with mandatory off-base training assignments and maneuvers, frequent six-month or longer deployments, and the anxiety and stress associated with deployments. Opportunities for extra-marital affairs abound and represent a constant threat for both partners with one or the other frequently out of the home. As our awareness of this lifestyle grows, the need to support our troops places an increasing demand on Christians. This is because they are both our neighbours and the defenders of our freedom. Thus, they deserve a double portion of our *"Love your neighbour as yourself"* (Mark 12:31) theology.

Finally, to prepare you for service in God's tactical unit as a spiritual warrior carrying the Good News to others and to raise awareness in Christian Canadian civilian gatherings about military life, I offer this advice: the most beneficial means of service and support for our troops is through an individually tailored approach. Author Dianne Collier agrees, as she stated in a private interview:

> And so my focus is on individuals. I will not take on any more projects …I've been told nobody else does what you do—but anybody can, I don't have a monopoly on the troops [or their families]. But nobody will take the time or make the commitment…so my focus is helping where I can—individually.[20]

There are four essentials to accomplishing a one-on-one approach. Collier spoke of two: time and commitment; there is no need to elaborate on them. I would add two more: to understand a family's lifestyle and to build relationships in that lifestyle. Every family unit, military or civilian, has unique and specific requirements; we cannot promote one-size-fits-all programs. Instead, we must begin by trying to understand the lifestyle to raise our awareness of its stress factors. We should also maintain a degree of familiarity with individual families and their situation. This will enable us to know what, when, where, and how to offer assistance. Assistance can be as simple as loading or unloading furniture or buying a pizza for lunch on moving day, watching a pet or keeping an eye on the house while families vacation. I maintain Christian civilians would quickly find effective areas in which to serve those who serve us, if they took the time, made a commitment, sought an understanding of military life, and built relationships with individual families.

---

[20] Dianne Collier. 'An Interview with the Author'. Interviewed at her home in Petawawa, ON. Taped and transcribed with permission. Date: January 20, 2009. All Interview quotes used with Author's consent—see transcript page 20.

Now we are prepared to move to a brief discussion of biblical parallels applicable to this chapter.

Biblical Parallels

I believe if we are honest with ourselves, we can find parallels between biblical characters and our own personal lives. Can you identify a biblical character that parallels your life? This lesson began with a look at me; can you see any biblical characters parallel to my life? Allow me to share some that I see in my story. In my youth, I was just as rebellious as the prophet Jonah. God called both of us to walk in His way and we both ran off in our own direction. Thankfully, I was not vomited up on the seashore by a big fish like Jonah. Once properly focused, my faith and Daniel's have certain parallels. Now, I was not about to step into a lion's den to prove my faith, but certainly my ministry path has been a day-to-day walk trusting the same God as Daniel trusted. Finally, there are parallels to my ministry and the ministry of Paul. Some scholars suggest he may have been divorced; regardless, Paul did serve as a roaming preacher with seemingly limitless faith. None of these are exact parallels, in fact they may simply be *"a poor reflection as in a mirror"* (1 Corinthians 13:12); but these parallels serve to encourage me in my ministry for God. In Scripture, I see God using the obedient and the not so obedient as well as the faithful and the not so faithful to accomplish His purposes. I have been in all those places.

Looking at the Christian civilian community around Pembroke, I discover two main parallels. The twelve spies who feared the powerful people of Canaan are like leaders of the Pembroke area Christian communities where some members have felt intimidated by the military. Just as Caleb and Joshua were ready to take on the challenge, these local leaders are willing to cooperate to reach an equally powerful people. Another appropriate parallel exists between the disciples of Jesus after Pentecost and the Christian community leaders. Both groups are part of God's tactical unit, drawn together by the Holy Spirit's power—warriors gifted for witness throughout the earth to the last hour (paraphrase of Acts 1:8). Finally, the military lifestyle parallels with Abraham, Jesus, Paul and numerous others. These will receive more attention in the subsequent biblical parallels section of each chapter.

This finalizes the introductory class on troop training. The next chapter turns to biblical and historic mandates for Christian engagement with our military. The poem introducing this class demonstrates that reviewing history is not unusual and the process should never be ignored.

Mapping Past Strategies
TWO

DETERMINING OUR HISTORY
Now history is not a subject we can afford to ignore;
In so doing, on other's mistakes we close the door.
Not to mention the horror of neglecting our own past,
For committed sins not remembered can recur too fast.
In Canada, our history does not extend back very far;
But Israel's goes beyond Dead Sea Scrolls in a clay jar.
We read old Abraham is where their history all began,
Then to Isaac, then to Jacob and on the tribe it did fan.
When famine did abound, the clan to Egypt did flee;
And slaves they wound up until Moses set the nation free.
The Red Sea this multitude crossed over as on dry land,
But Pharaoh's army it did sweep away at God's command.
To the wilderness they wandered to grumble and complain,
With miracles galore, their hearts still hard as that of Cain.
While God the Ten Commandments He did on Moses bestow,
They worshiped and praised the Golden Calf idol down below.
After scouting the Promised Land, rebellious refused to enter in;
So off they went to wander forty years in the desert for their sin.
At the appointed time Joshua led the nation to take the land,
But still rebellious and disobedient to God would be this band.
Through Judges, Kings and beyond, the history follows this way;
Thinking only of now and self, ignoring the past they did stray.[21]

---

[21] To fully understand Israel's history requires reading the Old Testament guided by the Holy Spirit but overviews of their history are found in Psalms 78, 105 & 106

The challenge everyone faces is what decisions should we make?
Do we learn from the past or continue in an old former mistake.
It is a free will choice to wallow deep in this world's sin,
Or to honour and praise the Creator from deep down within.
We paused to consider the unknown starting point for life,
And the human need caused by our own personal strife,
On the life a Christian should expect, to this I did preach.
World history, and yours too, now have lessons to teach.
Then we're on to sneak a peek thru boot camp's door.
A warrior God as our Saviour we'll discover and adore.

MAPPING PAST STRATEGIES

Each day, we represent history in the making just as our forefathers did before us. Due to this fact, it is important to reflect on their achievements and learn from them. This book proposes Christian Canadian civilian communities consider an intentional military ministry for our troops. When one takes up the challenge for this type of proposal, God's Word must form the foundation. Instead of developing an in-depth theology of military ministry, this book echoes biblical, theological, and historical patterns of involvement by God's people in ministry to our military and their families. As demonstrated in the preceding chapter, interest in military ministry currently abounds. By exploring history, we have a starting point for wrestling with reason and emotion to help resolve concerns about the validity of military ministry. We begin our exploration in Scripture at Genesis and touch highlights to the end of World War II.

Old Testament & Our Warrior God

God created; and the first couple promptly disobeyed their only commandment to *"not eat from the tree of the knowledge of good and evil"* (Genesis 2:17). Some may see the first couple's action as a simple act of disobedience, but it was in fact an act of rebellion against God. Adam and Eve desired to know as God knew; the devil said they could *"be like God, knowing good and evil"* (Genesis 3:5). As a result, human beings have lived in rebellion to God ever since. Sin was conceived in the heart of man as the first family fell short of God's expectations, His standard of excellence, His holiness. In the second generation, sin quickly grew into anger, envy, and hatred, resulting in Cain's premeditated murder of his brother Abel (Genesis 4). The founding family and their offspring continued in a downward spiral; finally, God said, *"I am grieved that I have made them"* (Genesis 6:7). Providentially, His love would not allow for the total destruction of His creation. *"Noah found favor in the eyes of the Lord"* (Genesis 6:8); thus, God kept for Himself this one man and his family as a remnant. *"Noah was a righteous man, blameless among the people of his time, and he walked with God. Noah had three sons...* [and God told Noah] *'...you will enter the ark—you and your sons and your wife and your sons' wives with you'"* (Genesis 6:9, 10, & 18). After the flood, God ordained a covenant with Noah and every living thing. He declared, *"Never again will the waters become a flood to destroy all life. Whenever the rainbow appears in the clouds, I will see it and remember the everlasting covenant between God and all living creatures of every kind on the earth"* (Genesis 9:15b–16). However, our sinful nature persisted. Fortunately for humanity, God's ultimate plan to deal with our stubborn, hard hearts was a perfect sacrifice of His Son, Jesus Christ, the Lamb of God. To begin a family tree for the coming Christ or Messiah, God called Abram. Scripture says of him: *"Abram believed the Lord, and he credited it to him as righteousness"* (Genesis 15:6).

The family line of Abram (later Abraham), Isaac (his son), and Jacob (or Israel—his grandson) constantly relocated. Abram commanded the first army of God's people. When his nephew Lot was taken prisoner, *"Abram...called out the 318 trained men born in his household and went in pursuit..."* (Genesis 14:14). Commanding Officer (CO) Abram devised a new military strategy. *"During the night Abram divided his men to attack them and he routed them, pursing them as far as Hobah, north of Damascus"* (Genesis 14:15). Later, we find Abram's victory resulted from God's direct intervention on his behalf. Melchizedek, *"king of Salem and priest of God Most High"* (Hebrews 7:1), proclaims: *"Blessed be Abram by God Most High, Creator of heaven and earth. And blessed be God Most High, who delivered your enemies into your hand"* (Genesis 14:19–20). In a subsequent incident, the song of Moses and Miriam portrays God as the direct cause of Israel's defeat of Egypt. *"The horse and its rider he* has hurled into the sea… *The Lord is a warrior…Pharaoh's chariots and his army he has hurled into the sea…Your right hand, O Lord, was majestic in power. Your right hand, O Lord, shattered the enemy"* (Exodus 15:1–6). The song continues in this pattern for twelve more verses and makes it clear that God was viewed as the direct cause of the Egyptians' defeat. Moses confirms this and commands parents to tell their children, *"We were slaves of Pharaoh in Egypt, but the Lord brought us out of Egypt with a mighty hand"* (Deuteronomy 6:21). For Moses, freedom came for the Israelites without them lifting a hand; God alone defeated the Egyptian army.

According to the Psalms, victory in the Promised Land was attributed directly to God:

We have heard with our ears, O God; our fathers have told us what you did in their days, in days long ago. With your hand you drove out the nations and planted our fathers; you crushed the peoples and made our fathers flourish. It was not by their sword that they won the land, nor did their arm bring them victory; it was your right hand, your arm, and the light of your face, for you loved them. (Psalm 44:1–3)

Certainly, the book of Joshua portrays the army of Israel as the contributing cause of defeat in conquering the Promised Land, but the poetry of the Psalms describes God as the agent who gave victory to the nation of Israel.

When Jericho was attacked following the forty years of wilderness wandering, God played a central role and positioned Himself between the armed guard (or front line) and rear guard (or home front). But, how does this connect with military ministry on the home front of today? To answer the question we must draw some assumptions. In Joshua 1:12–14 we read that the entire nation of Israel crossed the Jordan when Jericho was attacked, except for the wives, children, and livestock of Reuben, Gad, and half the tribe of Manasseh. Thus, it is reasonable to assume the rear guard was comprised of the women, children, elderly, and even livestock of the other tribes of Israel. In other words, as Joshua and the armed guard moved out followed by the presence of Almighty God, the home front of today formed the rear guard or rear flank. If we flip back to Exodus, we read of another battle, this one between the Israelites and Amalekites. Moses, Aaron, and Hur stood on top of a hill as the army of Israel, under the command of Joshua, fought the Amalekites. *"As long as Moses held up his hands, the Israelites were winning"* (Exodus 17:11), but when he tired and dropped his arms, the tide turned against Israel. Aaron and Hur were there to support the arms of Moses and in so doing gave Israel the victory. Joshua served as CO of the armed guard (on the front line) but did Moses, Aaron, and Hur not represent the rear guard (the home front)? This account may well be the first example of a front-line/home-front dichotomy, although not acknowledged as such in Scripture. If this were not an example of the front line supported by the home front, the trio of God's representatives would be symbolic of chaplaincy ministry. It then represents an account of the first example of a front-line/chaplaincy dichotomy. Even if these two possibilities are discounted, it demonstrates a theme of human forces utilized to fulfill God's plan.

There is one final passage of a front-line/home-front dichotomy to consider. In First Samuel, chapter thirty provides an account of a situation involving David and the Amalekites; without detailing the conflict, a key to home-front support is provided because of this encounter. David left two hundred of his six hundred men at the Besor Ravine with supplies. These *"men were too exhausted to cross the ravine"* (30:10). Upon their return, the armed guard cried, *"Because they did not go out with us, we will not share with them the plunder"* (30:22). However, David responded, *"No, my brothers, you must not do that with what the Lord has given us...The share of the man who stayed with the supplies is to be the same as that of him who went down to the battle. All will share alike"* (30:23–24). David proclaimed this as an everlasting statute and ordinance in Israel. In this text, the men staying with the supplies were the rear guard; if the battle turned against those on the front line, supplies and reinforcements were just across the ravine. One can view those not actively engaged in combat as a rear guard or home front, but regardless of what we call them, victory is a joint effort. All share alike today from the sacrifices made and victories won by those who have gone to battle for Canada. They paid a price we can never take for granted. In addition, we owe them our ongoing support by honouring veterans and standing behind today's troops. Finally, we must remember that God includes His people in battles on the front line and at the home front. He is either the contributory cause or an agent of defeat. Regardless, God remains on His throne in control of every situation in life—military or otherwise.

New Testament & Our Spiritual Warrior

In Luke 4:18–19 Jesus proclaims, *"The Spirit of the Lord is on me, because he has anointed me to preach good news to the poor. He has sent me to proclaim freedom for the prisoners and recovery of sight for the blind, to release the oppressed, to proclaim the year of the Lord's favor."*[22] Immediately after, Luke 4:21 records Jesus as stating: *"Today this scripture is fulfilled in your hearing."* Jesus was not referring solely to the financially poor, those locked in prison, the physically blind, or people oppressed by the Roman occupation, although these were part of His ministry. Jesus was speaking of His spiritual battle to set the souls of lost humanity free from this form of satanic bondage. His birth, life, death, burial, resurrection, ascension, and outpouring of the Holy Spirit were a declaration of war on evil forces in spiritual realms. In other words, Jesus took the war against evil to the highest possible level. Hadley and Richards put His struggle into perspective, commenting: "Anywhere militant evil exists, righteousness must marshall its forces to withstand, conquer, or be conquered. Jesus therefore was hardly a non-combatant."[23] Jesus Christ waged spiritual war against forces of evil to liberate humanity as surely as Abram waged temporal war against the Amalekites to rescue Lot. Both marshalled their forces to withstand and conquer the militant evil threatening to conquer their loved ones. Jesus was a warrior! As well as being so much more: priest and sacrifice, prophet, teacher, healer, miracle worker, and minister to all—including military personnel.

---

[22] Also see Isaiah 61:1–2
[23] Donald W. Hadley and Gerald T. Richards. *Ministry with the Military: A Guide for Churches and Chaplains*. Grand Rapids: Baker Book House, 1992. 39

In the Lord's Prayer, we petition God, asking His *"will be done on earth as it is in heaven"* (Matthew 6:10). In the Old Testament, God frequently used His people to accomplish His plan or purpose. Since we serve an unchanging God, it would not be unrealistic to expect Him to continue to use His people to ensure His will is done on earth as it is in Heaven. Jesus instructed us: *"So in everything, do to others what you would have them do to you, for this sums up the Law and the Prophets"* (Matthew 7:12). Other passages to consider in seeking a New Testament biblical mandate for evangelism and ministry to our troops can be found in Matthew 28:19–20: *"Therefore go and make disciples of all nations, baptizing them in the name of the Father and of the Son and of the Holy Spirit, and teaching them to obey everything I have commanded you."* Also, Acts 1:8: *"But you will receive power when the Holy Spirit comes on you; and you will be my witnesses in Jerusalem, and in all Judea and Samaria, and to the ends of the earth."* These two passages are versions of the Great Commission calling us to witness and make disciples of all—including those in the CF. Finally, the above passages can be summed up by the Great Commandment:

> *One of the teachers of the law…asked him, "Of all the commandments, which is the most important?"*
>
> *"The most important one," answered Jesus, "is this: 'Hear, O Israel, the Lord our God, the Lord is one. Love the Lord your God with all your heart and with all your soul and with all your mind and with all your strength.' The second is this: 'Love your neighbor as yourself.' There is no commandment greater than these."* (Mark 12:28–31)

The basic commandment can be reduced to: Love God—Love Neighbour. It is the call for all Christians to treat all people as they would wish to be treated. How? By doing His will on earth as it is done in Heaven. His will involves sharing the Good News of salvation, the Great Commission. As Christians, for us to do anything less is a direct violation of the Great Commandment, which calls us to love our neighbour. As an act of love, we should share the Good News of salvation with our neighbours. Families on base are our neighbours as much as our spouse, children, the folks next door, or an orphan we support in a foreign country. It may be something we approach with fear and trembling, but God calls us to advance as members of His tactical unit and deliver the Good News to our neighbours in the military.

Jesus was the first, but certainly not the last, to proclaim good news to the poor, imprisoned, blind, oppressed, and even the military. In fact, a Roman centurion preached in his own style. Jesus said about him, *"I tell you the truth, I have not found anyone in Israel with such great faith"* (Matthew 8:10). The visible faith of this centurion apparently said more than the preaching of religious leaders of the day. There is a proverb applicable here: "Actions speak louder than words." Christ was saying to the religious leaders of His day, "Your actions speak so loudly to people they cannot hear a word you are saying." He commended the centurion's actions because they spoke so loudly he did not need to utter a word. Moving past the day of Pentecost, we read of Peter making a trip to visit and share the Good News with *"...Cornelius, a centurion in what was known as the Italian Regiment"* (Acts 10:1). Paul, too, was never shy about sharing the Word. He shared the Good News of salvation with his Philippian jailer, a member of the Roman army, in Acts 16:25–34. Later, in his letter to the Philippians, Paul writes, *"...it has become clear throughout the whole palace guard and to everyone else that I am in chains for Christ"* (Philippians 1:13). Paul may well have had more opportunities to witness to military personnel than all of the other apostles combined. As Christians, we are disciples or followers of Christ Jesus. He is our Spiritual Warrior! As a result, we have a mandate to follow His example. Christ shared the Gospel universally, although primarily to the Jewish people. Christians, today, are called to continue the tradition as did Peter and Paul. They carried the Good News to Cornelius, the Philippian jailer, and unnamed others in the military during their ministry in early church history.

Canada & War—Historically

Within its short history, Canada has experienced few military skirmishes on home soil. Canadians familiar with our history will recall early settlers battled with First Nations people; the 1759–60 French/English battle for Quebec; the attempted invasion by the Americans in 1812–14; and

Louis Riel's rebellion ending in 1885. There were also lesser-known struggles in 1775, 1816, 1837 and 1866.[24]

However, we start with the War of 1812, which impacted Canadian church history as Murphy notes:

> By the end of the War of 1812 [the Catholic Church] had managed to avert the threat that had been hanging over it since 1793. In the course of the war, Bishop Plessis and his clergy had urged the Canadiens to be faithful subjects of the king. A number of Canadiens had taken part in the fighting, and some priests had served as chaplains in the army...As a reward for his loyalty during the conflict, in 1818 the British government had recognized Plessis as the 'bishop of the Roman Catholic Church of Quebec', with the right to sit in the Legislative Council.[25]

Protestant traditions were also influenced by this war:

> The spectacular expansion of popular evangelicalism was abruptly curtailed, especially in Upper Canada, by the War of 1812...Baptist congregations were permanently dispersed...the very life-line of early Methodism was cut off when the Upper Canadian border was closed to all American denominations.[26]

Thus, war brought legitimacy to the Catholic Church, but slowed growth of protestant evangelicalism in our nation. In addition, the War of 1812 established the foundation of chaplaincy ministry in what is now Canada.

---

[24] John Robert Colombo and Michael Richardson, compiled by. *We Stand On Guard: Poems and Songs of Canadians in Battle*. Toronto, ON: Doubleday Canada, 1985. Accounts of these lesser known struggles 37, 50, 53, 61

[25] Terrence Murphy, ed. *A Concise History of Christianity in Canada*. Don Mills, ON: Oxford University Press, 1996. 88

[26] George A. Rawlyk, ed. *The Canadian Protestant Experience 1760–1990*. Kingston, ON; McGill-Queen's University Press, 1990. 41

Canada's participation in the Great War of 1914–18 changed our history forever. Colombo provides a picturesque description of the change:

> Canada's participation in the First World War marked its coming of age.
> "There they stood on Vimy Ridge," orated Lord Byng. "Men from Quebec stood shoulder to shoulder with men from Ontario, men from the Maritimes with men from British Columbia, and there was forged a nation tempered by the fires of sacrifice and hammered on the anvil of high adventure." The troops crossed the Atlantic as colonials, and re-crossed it as Canadians.[27]

CBC offers a basic account in a Remembrance Day article: "In many ways, the identity of the young country was forged on those bloody battlefields."[28] Those bloody battlefields include Vimy Ridge, Hill 70, Passchendaele, and Ypres. We should remember almost 620,000 Canadians served in World War I, and of these over 66,000 did not live to return home—in excess of ten percent. The Christian church in Canada did not take a pacifist stand toward the Great War, but rather played a significant role in the "war to end all war."[29] The book *Conscription 1917* is a collection of essays—one of which is especially applicable to this chapter: "The Methodist Church and World War I."[30] The author writes:

> Few Canadian Methodists worried about war in the spring of 1914.
> In the autumn of 1914, however, church leaders abandoned their critical acquiescence in war in favor of an unquestioning

---

[27] Colombo. 7
[28] http://www.cbc.ca/canada/story/2008/11/07/f-remebrance-day.html; accessed February 21, 2013.
[29] Murphy. 335
[30] Ramsay Cook, Craig Brown, Carl Berger, editors. *Conscription 1917: Canadian Historical Readings*. A selection of articles from the *Canadian Historic Review* and other volumes. Toronto: University of Toronto Press, n.d. Essay by J.M. Bliss: "The Methodist Church and World War I". Reprinted from *Canadian Historical Review*, XLIX (3), September, 1968.

belief in the righteousness of the conflict and the church's duty to play a positive role in achieving victory.[31]

Bliss goes on to note: "Roughly 90 per cent of the Methodist clergy who served were combatants. This remarkably high figure illustrates the intensity of Methodist support and verifies the assumption that the clergy saw themselves as fighting a holy war."[32] We see a series of developments occurring in Methodist leadership as this war dragged on. First was the shift in thinking in 1914 mentioned above. By "the autumn of 1916 Methodist leaders began to call for conscription."[33] The author goes on to note that by 1918, "Methodists decided that pacifism was no longer a legitimate Christian doctrine."[34] Based on actions of Methodist leadership from 1914 to 1918, it is apparent Methodism in the past took a pro-war stand and supported the military. To paraphrase Hadley and Richards: Methodists realized a militant evil existed and righteousness needed to marshall its forces to withstand and conquer, or be conquered. Methodists therefore could hardly be seen as non-combatants.

Among churches in Canada, Methodists were not alone in their pro-war stand. Murphy writes of this as a general trend among Protestants:

> Protestant clergy preached sermons urging young men to do their patriotic and religious duty by enlisting in the army; they threw their support behind conscription when the government introduced it in mid-1917; and, in the election that followed, they called on Canadians to vote for the wartime Union government in the name of God and Country. Canadian Protestants were both intensely loyal to Britain and fervently nationalistic.[35]

---

[31] Cook. 39 & 40
[32] Cook. 43—with a reference to: *Journal of Proceedings of the Tenth General Conference of the Methodist Church* (Toronto, 1918), pp.140–1
[33] Cook. 46
[34] Cook. 49
[35] Murphy. 335—with a reference to Michael Bliss, 'The Methodist Church and World War I', in Carl Berger, ed., *Conscription* 1917 (Toronto: University of Toronto Press, n.d.), 43–4, 46–7

On the other hand, Canadian Catholics were split. The francophone population viewed "the war as a European issue with no bearing on North America"[36]; but a study indicated that "English-speaking Catholics in Toronto viewed the war in much the same light as their Protestant fellow citizens…"[37] In addition, Grant notes: "In 1914…the churches rallied without hesitation to the national cause …Churches that had had so large a part in establishing the norms of Canadian life could not be indifferent to an enterprise that involved the nation so deeply."[38] Rawlyk adds his comments to the dialogue:

> The churches' participation in the war effort was total. Chaplains were recruited for overseas service and pastoral duties were enlarged to cope with the trauma of war at home. Sermons were written to evoke commitment, sacrifice, and honour, and the Gospel assumed a decidedly muscular air.[39]

Today, facing the evils of terrorism, it is interesting to see trauma acknowledged as a home-front issue in WWI. Commitment and sacrifice were emphasized, and chaplaincy ministry flourished. The War of 1812 brought adjustments to Christian traditions, but the First World War focused on commitment to victory over evil. In light of this, Canada could hardly be viewed as a non-combatant.

When World War II began, conscription was no longer an issue the church had to deal with. As noted below:

> Canada declared war on Germany on September 10, 1939… Compulsory service for home defence began in June 1940. This required a nationwide registration of everyone over the age of 16, so that the government could direct both national military service and any civilian labour that was related to

---

[36] Murphy. 336
[37] Murphy. 337
[38] John Webster Grant. *The Church in the Canadian Era*. Burlington, ON: Welch Publishing Company Inc, 1988. 113
[39] Rawlyk. 143

the war effort. In 1942, compulsory overseas service was introduced…[40]

It was the national government who adopted the earlier comment by Hadley and Richards by recognizing a militant evil existed. This evil called for the nation to marshall its forces to withstand, conquer, or be conquered. Canada became a combatant in WWII. Rawlyk writes: "The day-to-day role of clergy during the Second World War was much the same as it had been in the first: chaplains were dispatched and pastoral duties were once again enlarged to encompass the task of providing comfort on the home front."[41] Murphy adds this comment: "…the churches rallied to support the war effort, but with little of the fervour they had shown in 1914."[42] Grant expounds in more detail on church participation:

> The churches adapted themselves to the peculiar requirements of a state of war. They cooperated in the appointment of service chaplains, of whom…almost all won the respect of the men and women whom they served. Priests and ministers at home brought comfort to survivors and to those who anxiously awaited word of friends and relatives in Britain or occupied Europe.
>
> So far the role of the church was much what it had been in the first world war, but there the resemblance ended. Emotional appeals…were replaced by sober determination to finish a messy but necessary job… They moderated their protestations of loyalty, refrained from overt participation in the war effort, and sometimes prayed for national enemies.[43]

There was a need that the government of Canada was determined to meet even if the church role was somewhat limited.

---

[40] Janice Summerby. *Native Soldiers—Foreign Battlefields*. Remembrance Series, Publication of Ministry of Veteran Affairs, Canada, 2005. 21
[41] Rawlyk. 189
[42] Murphy. 354
[43] Grant. 150–51

In the chaplaincy program, there was also a need. It was excessive enough that exceptions to official policy were allowed. Davis, a chaplain with the artillery, shares from his experience:

> In 1939, I was only six weeks out of Seminary when I was sent to the parish of Petawawa...
> When war broke out in September, I was asked...when I would be joining the active forces...I was moved to investigate the process...I was somewhat dismayed...when I discovered one of the requirements of entering the chaplain service as a padre was five years experience in parish work.[44]

He goes on to write of a visit to the parish by the chaplain general, who later returned to tell him this was a young man's war and ask why he was not in uniform. Davis adds, "I sent in my application right away and was accepted."[45] In this case, clearly a lack of chaplains superseded the official requirements. As a point of interest, the prerequisite in 2012 for "parish work" to enrol in chaplaincy is two years.

Canada no longer depends on conscription to meet its defence requirements. We are blessed by men and women who step forward and answer the call to serve our country as military professionals. The church does maintain a presence in the military through chaplaincy; however, we cannot overlook the para-church organization MCF. The CF Chaplain Branch is an extension of the church. The Handbook of the Interfaith Committee on Canadian Military Chaplaincy (ICCMC) makes the extension clear in its opening paragraph. Chapter II – Principles of Canadian Military Chaplaincy:

> Historically, the Chaplain Service in the Canadian Forces has been rendered by a partnership between the Government of Canada, as represented by the Minister of National Defence, and the Canadian Conference of Catholic Bishops and the

---

[44] Eldon S. Davis. *An Awesome Silence: A Gunner Padre's Journey Through The Valley Of The Shadow*. Carp, ON: Creative Bound Inc, 1991. 9
[45] Davis. 10

Canadian Council of Churches. With the evolution of the Service the partnership included churches and faiths beyond the foundational agreements, in order to reflect the diversity of Canadian Society and to meet the needs of all in the Canadian Forces.[46]

Chaplains are an essential, integral part of our military who deserve and need our prayers and support.

Chaplains, more than any group, must reconcile in their own minds the obvious opposing Christian views of life and war resulting in death. An historical novel, *In This Sign*, is written by "a retired chaplain with the Canadian Armed Forces. Based on true events…"[47] Although Ralph is a fictional World War II character in the novel, there is no doubt his struggle to reconcile war and Christianity is not unique to him. In a Christmas Eve message, Ralph offers the men these words: "You may wonder if you serve a just cause. You may have difficulty rationalizing the killing in battle. Does not the sixth commandment say, 'Thou shalt not kill?' I was taught a wider interpretation of the Hebrew words to mean, 'Thou shalt not murder.'[48] It would sound as if he has settled the matter in his own mind. But, later he continues to wrestle with the question and as he does Ralph is faced with the reality of a life and death situation:

> Reason and emotion wrestled within him.
>
> "We are battling an evil enemy. If we don't, who will? What does the Commandment 'Thou shalt not kill' really mean? How can one justify the taking of another man's life? Soldiers are taught that their very existence is predicated upon the need to kill the enemy." Ralph squirmed in his seat, thinking. "This whole question places the Christian between a rock and a hard place. I must try to resolve this dilemma or go crazy."

---

[46] http://www.cmp.cpm.forces.gc.ca/cfcb-bsafc/pub/iccmch-gciamc-eng.asp, accessed February 21, 2013

[47] Reverend Lyman R. Coleman Lt. Co. (ret.). *In this Sign: An Historical Novel.* Renfrew, ON: General Store Publishing House, 2005. b.c.

[48] Coleman. 100

Precisely at that moment, three German soldiers leapt onto the road from behind a large bolder. The first man raised his hand to halt the vehicle…the German soldiers…immediately raised their weapons to bear. The pipe major… opened fire…All three Germans fell to the ground…

…their faces contorted in death.[49]

For Ralph, the driver, and the pipe major sitting in the back seat, it became a choice—kill or be killed. It was just as Hadley and Richard claimed: in front of them a militant evil existed and righteous or not they needed to marshall their forces to conquer, or be conquered. This Anglican priest therefore became a combatant, not using a weapon, but by association. In the same way, the decision about ministry to our military is a choice individual Christians will make; there is no right or wrong answer. But we cannot make an informed decision until we first wrestle with reason and emotion as Ralph was in the process of doing. I offer this biblical, theological, and historic overview to stimulate the reasoning and emotional wrestling process in advance of such a decision. We must remember that Canada, historically, has not been a violent, aggressive nation; rather, we are respected internationally for our role as peacekeepers. However, neither is Canada a pacifist nation and many Christian churches in Canada have historically encouraged and endorsed use of our military. Therefore, the Christian church is involved in the military by association, just like Ralph. It encouraged defending our border in 1812, it encouraged assisting our allies in past world conflicts, and it continues to profess support for our troops today.

Finally, we have reviewed a great deal of material covering a massive span of history, so it seems prudent to offer a brief summary. In the beginning, God created; and through rebellion humanity messed up creation in disobedience to His one command. With the introduction of sin, society started on a downward spiral leading to murder and war. Since the beginning of time, there has been a spiritual battle ongoing, although it was considered temporal warfare by the patriarchs and their

---

[49] Coleman. 157–8

offspring as they marched to the Promised Land, then into captivity, and back to being an occupied nation. Spiritual warfare intensified with the arrival of Christ, God's only Son. Herod's attack on children in an effort to destroy baby Jesus, the Devil's wilderness temptation, and the Cross were attempts by evil forces to foil Christ's earthly mission. Paul, too, battled spiritually, and all Christians should expect to be attacked by the enemy of their soul. The Bible clearly calls God's people to the front line of battle—it may be a personal spiritual battle, or defending the rights of widows and orphans, or protecting justice, or standing to conquer militant evil. Psalm 82 inquires: *"How long will you defend the unjust and show partiality to the wicked?"* It next offers instruction: *"Defend the cause of the weak and fatherless; maintain the rights of the poor and oppressed. Rescue the weak and needy; deliver them from the hand of the wicked"* (82:2–4). Jesus echoes this in the Parable of the Sheep and the Goats. Here, Jesus reminds us: *"I tell you the truth, whatever you did for one of the least of these brothers of mine, you did for me"* (Matthew 25:40). We are not only called by God to fight evil, we are also called by our nation. Historically, Canadians have participated as combatants in self-defense, stood for justice for all, and served as peacekeepers to support struggles against injustice, evil, and oppression. Biblically and historically, members of God's tactical unit have not taken a pacifist stand in terms of spiritual warfare. Is the root of all warfare spiritual? In Canada, we no longer depend on conscription to meet our defence requirements. Thus, Christian Canadians are free to reconcile this question as it relates to temporal battles.

Historic Theological Foundations

The book by Hadley and Richards is dated by virtue of its 1992 publishing, but it remains a very worthwhile resource for military ministry. I am indebted to them for their comprehensive work in this field of ministry. The authors remind readers of five basic theological principles, summarized below:

First, consider whether God is calling you to a specific ministry to the military. We are all called to ministry in general, but some are called to a specific ministry.

Second, consider the Luke 12:39–40 concept of the thief in the night; this is often how death comes. Due to the high risk factor in military life, Christians should have a sense of urgency in sharing the gospel to thwart the thief.

Third is the generally held concept of churches being a loving family and God's children. The image is meaningful to individuals and families separated from their biological family. They need the love and support of a welcoming church.

Fourth, a thought that ties into number three, the authors point to the emphasis of our Christian faith based on love. Generally speaking, military life is rough, tough, depersonalized, and demanding; Christians can counter with warm-heartedness, sincerity, understanding, and caring. In this, we demonstrate the love of God.

Fifth and finally, consider these lost sheep as they wander the globe in a Canadian military uniform. Both the wanderer and their family need to be shepherded through our Lord's chaplains, pastors, and others called to this ministry.[50]

These are additional thoughts to consider when a congregation is exploring the establishment of an intentional military ministry.

Biblical Parallels

This chapter has examined Scripture and history, particularly in terms of war and chaplaincy or the place of God on the temporal battlefield. It has included a brief review specifically focused on the nation of Canada, for which no biblical parallel is possible. In terms of Old Testament summaries, the poem opening this chapter pointed to Psalms 78, 105 and 106, which include a review of Israel's history, but many Old Testament passages contain partial reviews of Israel's past. In addition, this sub-section is partially paralleled by Stephen in Acts 7:1–53, Paul in 1 Corinthians 10:1–13, and to some extent throughout the book of Hebrews, especially chapter eleven. There are numerous other short references to the Old Testament within New Testament writings—particular the letters of Paul. In fact, enough echoes from the past to

---

[50] Hadley. 41–3. Summary of their work

prompt Richard B. Hays to write the book *Echoes of Scripture in the Letters of Paul*.[51]

The life of Jesus is recounted in the four Gospels and a number of other New Testament passages. Those of specific interest include Acts 3:11 to 4:22, Acts 5:17–42, 1 Corinthians 15:1–58, and Philippians 2:5–11. Some of these accounts include Old Testament summary material leading up to His birth, life, death, burial, resurrection, ascension, and outpouring of the Holy Spirit. The book of Isaiah is an excellent example of this. In addition, Paul's personal history is often retold. His testimony begins with Luke's description in Acts 9:1–19, followed by accounts in Acts 22:3–16 and 26:9–18. Paul writes of his conversion in Galatians1:11–24 and his hardships in 2 Corinthians 6:3–10 and 11:16–33 as two examples.

This concludes our second introductory lesson on troop training. These first two lessons were offered to establish a foundation on which to present various lifestyle issues found in the military; and ultimately, to parallel these with biblical similarities. They have also attempted to answer the four assumed questions raised in the opening paragraph. With heartfelt gratitude, I also acknowledge those who have faithfully preserved the history of our Christian heritage in Canada; this book is more valuable because of their efforts.

We now turn our attention to the book's primary objective—preparing the troops of God's tactical unit. It is expected the awareness of the need for Canadians to support our troops will increase by drawing parallels between soldiers of Christ Jesus and their CF counterparts. May this begin in Christian civilian congregations! Thus, our lifestyle tour will begin with day one in military boot camp. However, before moving on, please pause and reflect on the introductory poem for boot camp. We must never forget Jesus hung on the Cross, died, rose from the grave, ascended to Heaven, and poured out the Holy Spirit for us as individuals. His grace is sufficient for every generation throughout history but we must individually accept His provision. If you have not already done so, consider committing your life to Christ and enter the first day

---

[51] Richard B. Hays. *Echoes Of Scripture In The Letters Of Paul.* New Haven, CT: Yale University Press, 1989.

of His boot camp. It is a camp focused on discipline, commitment, obedience, sacrifice, and training in weapon handling—*"the sword of the Spirit, which is the word of God"* (Ephesians 6:17). This is an opportune time to consider the reality of Jesus Christ as the Lord and Saviour of humanity. The poem is lengthy but deals with eternal salvation. It allows committed Christians an opportunity to reflect on their commitment and others to consider such a commitment to "Someone New—Just For You." Yes, He did live, suffer, and die only to be raised again from the dead—just for you!

Boot Camp
THREE

SOMEONE NEW—JUST FOR YOU
On an adventure I invite you to come along,
As I put this message into a rhythmic song.
But let me give the credit to Whom it is due;
T'was the Holy Spirit who guided this thru!
I'll try not to lull you to sleep with the tune;
Rest assured you will have it read very soon.

Once upon a long, long, long time ago,
To the prophet Isaiah, God did go.
Now to the prophet, Someone new God did explain;
Someone not just for Isaiah, but he did not complain.
Thus, of Someone very special the prophet wrote:
The Spirit is on me and anointed I am—take note!
Good News to the poor—I come to speak;
And to bind up broken hearts—I do seek.
For the captives to sin—freedom I bring,
To release from darkness allowing them to sing.
The year of the Lord's favour—I come to claim;
And the day of His vengeance to proclaim.
To those who mourn—comfort I extend;
And to provide for the grieving is my end.
On these a crown of beauty I will bestow,
With the oil of gladness they will overflow.

Then add a garment of praise to their attire—
Ashes, mourning and despair I will retire.[52]

Now who was the One of whom the prophet did cry?
A resident of his day? A man of History gone by?
Or was it on a far off hill that he did see—
The Messiah who bled for you and for me?
Was this One—a present, past, or future being?
Number three—the future—I am seeing.
But let's look at all three one at a time;
As we continue with this little rhyme.

Now look in the streets and you will soon find
No one of Isaiah's day on the prophecy to bind.
For Israel was overtaken with corruption and sin;
And pointing out errors, no friend a prophet can win.
We live in an age no different from that trend;
Don't fool yourself trying otherwise to pretend.
Of course the task of friend to all—was not Isaiah's to be!
Nor is it yours and mine—hope that you can see!
Our call is to turn hearts back to God;
This to some makes us seem strange and odd.
To accomplish his goal—things to come Isaiah foretold,
Which had the power to shake even the most bold.
Or he'd forth tell the past—thus sharing God's Word,
But too often it was not heeded or else not heard.
Thus, if what you desire is a challenging career,
Fulfill your call to witness; and make it quite clear!
Sharing the Word, like a bold prophet of old,
Is the call on our life by Jesus—we are told.
But since the prophecy is not found in Isaiah's time,
That means I will have to continue with my rhyme.
So I'm off into history seeking the pattern you see.
Is it back then or way beyond Isaiah in number three?

---

[52] Paraphrase of Isaiah 61:1–3 (NIV)—Also quoted by Jesus in Luke 4:18–19

Well if the prophesy is not about Isaiah's day—
Maybe it concerns God's Word from yesterday.
So I turned back the clock of time—to read:
In Genesis—of Abram—please take heed.
Four kings they came and attacked the town;
Found Abram's nephew and family of renown.
They seized them and carried them off in despair,
Brokenhearted captives to the prophecy compare.
But one did escape and to Abram he fled—
So with all of his men to their aid—Abram sped.
His strategy: divide the warriors, launch an attack,
And ruined the enemy to save the family pack.[53]
Just one man through the Spirit's power
Rescued his family and saved the hour.
Many of the things the prophet wrote about
This passage most certainly does loudly shout.
But not enough to make Abram its superstar.
No the prophecy's "Someone" is better by far.

Today the same Spirit's power we hold in our hearts;
Living water flowing from within is where it all starts.[54]
By this Jesus meant the Holy Spirit we would receive
After He returned to Glory; and we did believe.
Now you are the family member that Christ has set free.
Therefore, just like Abram of old you are called to be.
Start by gathering to yourself warriors in prayer,
The ones you can count on who will always be there.
Next draw up a strategy and prepare to attack!
In so doing, you'll be giving the Enemy a smack.
Finally with family and friends take time to share;
Thereby rushing to the aid of those for whom you care.
But Abram's temporal battle was just the beginning,
For it was of spiritual war the prophet was singing.

---

[53] Paraphrase of Genesis 14:11–16
[54] Paraphrase of John 7:38–39

A prophecy not for Isaiah's day
Or from the past in some strange way.
So to the future we must now look,
Because we have it all written in the Book.
Please focus your attention on Luke chapter four,[55]
And here see Jesus open the spiritual battle door.
Of the Isaiah passage, this He claimed to fulfill.
For humanity, God's PAID stamped on *our* sin bill.
From temporal battles in days gone by—
To spiritual war in earth and 'bove sky.
The war on God began with Satan's fall,
But Jesus was sent to end it once and for all.
An attack on his infant life was foiled;
As Herod attempted the joy to be spoiled.[56]
And Satan himself in the desert took his turn
To tempt Him to yield; thus in Hell with him burn.[57]
And although the Devil tempted in every way,
From sin our Saviour did willingly stay.[58]
As He drew near the Cross with antagonism all about,
He said, *"The prince of this world will be driven out."*[59]
But on the old rugged cross we all hung Him to die;
We would finally be rid of Him—and so good-by.
He admitted we were right as He hung on that Cross:
"It is finished,"[60] He said.—But His death was not a loss.
"It is finished" did not an end to His earthly life mean,
But victory over Death and the Grave—soon to be seen.
Thus, on the first day of the week some women found:
He had risen! Death and Hell could not keep Him bound.[61]
Then in the future, John describes even more we'll see:

---

[55] Read Luke 4:14–21
[56] Read Matthew 2:13–18
[57] Read Matthew 4:1–11
[58] Paraphrase of Hebrews 4:15
[59] John 12:31
[60] John 19:30
[61] Read Matthew 28:1–7; also read Acts 2:14–41

That of His coming again in power and with majesty.[62]
Just as Abram we were told his whole family he had saved,
Christ rescued our soul for all Eternity from sin depraved.
After ascending to Glory, the Holy Spirit was poured out;
Because of Him, we are empowered to witness all about.[63]
So when God spoke to Isaiah about Someone new,
It was Someone just for you and for me—it is true.
He came for each one—to die for their sin,
As if theirs was the only soul He had to win.

So what is required of all believers in Christ today?
"What law is most important?" the teacher did say.
The Great Commandment, Jesus would recall—
With Love God—Love Neighbour above all.[64]
It is a difficult chore, one must admit,
To love sinners and hate the sin they commit.
So rather than try to decide their right or wrong,
Simply follow Christ's direction and tag along.
He told us what to do in the simplest of form,
Serve others—even a glass of water fits His norm.[65]
To point this out—a parable He provided;
It involved the sheep and the goats divided.[66]
On His right stood the sheep so righteous and pure;
To His left the unrighteous goats without a cure.
A blessing given or ignored against—guess who?
Was the only fence that separated these two.
So who was blessed or ignored?
Surely not Jesus Christ the Lord?
No, t'was a sister or brother on the edge,
As they sat teetering upon some ledge.

---

[62] Read Revelation 19:11–16, plus chapters 21 and 22
[63] Paraphrase of Acts 1:8
[64] Paraphrase of Mark 12:28–34
[65] Paraphrase of Matthew 10:42
[66] Read Matthew 25:31–46; also read all of Isaiah 58—especially vv. 6–7

You know—the one in need of life's provision—
Food, water, housing, clothes, sick or in prison.
The sheep reached out with a helping hand to lend,
But the goats thinking of self had no time to spend.
   Thus you see—what was done for one of these
      Was done unto Him—if you please.
For where your treasure is, there your heart will be.[67]
   Is it focused on self or the human family tree?
Now you may think, "I'll make it in Heaven's door!"
   But what does God really have in store?
   We can only see the outward woman or man,
But God looks at the heart when making His plan.[68]
   So if Christ really is your first love true;
   You can count on the Holy Spirit the rest to do.[69]
He'll give you a task unimportant you might bet,
   But be faithful to it and the bigger they'll get.
Thus follow one, who not long ago passed away,
   Mother Theresa, like a sheep lived day by day.
One person at time she did serve in God's name;
  "The Lord is my Shepherd," she would acclaim.[70]
  A life like hers is a special exception—is it not?
Yes, because God does not cast us all in one lot.
   But her example shines like a star in the night
That points to the difference between left and right.
   Now the choice is up to you and in plain sight:
Will you be found on Christ's left or on His right?

Today's world is similar to Abram and Isaiah's time;
Still our family we can save according to this rhyme.
   Thus, in closing these thoughts let me recap,
   By relating the prophecy to water from a tap.

---

[67] Paraphrase of Matthew 6:21
[68] Read 1 Samuel 16:7; 1 Chronicles 28:9; 2 Chronicles 6:30; also read Acts 15:7–11
[69] Read Romans 8:1–17
[70] Read Psalm 23

In Isaiah's day sin and corruption had the valve closed tight,
  As these two do today for those without Christ in sight.
But from the past Abram showed for our family what to do:
Rally the troops, launch an attack, open the tap a turn or two.
  When Christ poured out the Spirit on all who believe,
  Water gushed from the well abundantly for all to receive.
    Now the tap is wide open and you have the power
  Through Christ to rescue your family and save the hour.
    The journey begins at the foot of the Cross;
  This is the point where you acknowledge a new Boss.
    Now if this is a step you have yet to take,
  Pray the prayer below for your soul's Eternal sake.
      No longer a slave to sin you will be;
    Since, in Christ, we are all set free.[71] AMEN.

A Prayer of Confession

Father God, I confess I have sinned against You. Thank You, Lord Jesus, for paying for my sins; please forgive me. I receive You as my personal Saviour; take my life and use me to confess Your name, to pray and to follow the leading of the Holy Spirit. As my first step of faith, I accept Your forgiveness here and now. Amen.

If you have prayed this prayer with a sincere heart, I urge you to seek a Bible-teaching church, speak to the pastor about your confession of faith, and become involved in the life of that church.

Boot Camp

The first chapter of any book represents its starting point, its birth; although it may have been conceived and allowed to develop in an author's mind for some time. Likewise, Christians remember and identify their initial commitment to Jesus Christ as their birth into the Christian faith. As with a book, their decision is often formed and well

---

[71] Paraphrase of Romans 8:1–4.

reasoned out in their mind and heart prior to this birth. Similarly, it is plausible to assume a Canadian Forces (CF) member could consider the first day of training as their birth into a new career. However, I have not experienced such a birth. We hope these recruits understand and have thought through the implications of their decision. Because of a lack of personal knowledge or relevant ministry accounts shared on this subject, the following quote is provided to give one soldier's introduction to the CF. Ogle identifies Bass as a recruit who:

> ...sat for hours in the uncomfortable, moulded plastic chairs...
>
> About two-thirty in the morning, a sergeant entered the terminal... quickly took control...and ushered [the recruits] onto buses...As the doors slammed shut...the sergeant finally spoke.
>
> "Do not speak unless you are spoken to. Remain seated until you are instructed otherwise. Do not sleep. Do not ask any questions. Do as you are told."
>
> With his cryptic message delivered, the sergeant stood for the rest of the trip, glaring at the passengers. The bus ground to a halt forty-five minutes later at the Canadian Forces Recruit School, Cornwallis, Nova Scotia.[72]

This quote offers readers a firsthand taste of boot camp from one person's experience. It is the birth date for his military career; but the challenges of the next few weeks will have a long-term impact on his and every other recruit's life. What is drilled into these young women and men will influence their attitude and character for the rest of their days here on earth. As the first lifestyle issue, we will focus on five foundational principles taught in boot camp; all are required both in the military and in training to be *"a good soldier of Christ Jesus"* (2 Timothy 2:3). They are not the only essentials recruits are taught, nor the only

---

[72] James Ogle and Darnell Bass. *What Manner Of Man: Darnell Bass and the Canadian Airborne Regiment*. Renfrew, ON: General Store Publishing House, 2006. 15–16 (Warning: contains coarse language.)

fundamentals for Christian members of God's tactical unit. However, they can be applied to both types of soldiering. These five common discipleship characteristics are discipline, commitment, obedience, sacrifice, and weapon handling.

In the opening quote taken from Ogle, all five discipleship qualities are portrayed military-style; can you identify them? Training begins day one! This is what I find in the passage. In part, discipline deals with self-control and becoming subject to the point of trusting an authority. Sitting on uncomfortable, molded plastic chairs until two-thirty in the morning waiting for a bus requires a great deal of self-control for most people, myself included. The sergeant is the authority speaking to these new recruits; any discomfort they have experienced is normal and can be expected. Along with this comes the unspoken implication that these recruits better be committed to their military career. To rephrase the statement, the sergeant, through his body language and terse commands, is saying, "If you can't cut this day, you better cut and run now because things are only going to get tougher." He is the first authority the recruits must learn to trust. How does he gain their trust? He begins by informing the recruits of exactly what will be expected of them. As a result, obedience is likely the most obvious of the five characteristics as the sergeant strips away all civilian privileges with his "Do not" commands and offers only one "Do" obligation—"Do as you are told." This echoes a claim by missionary J. Hudson Taylor: "The Great Commission is not an option to be considered, but a command to be obeyed."[73] Next, the sergeant's sacrifice must be read into the scene since it is more difficult to identify; he may be sacrificing his opinion on training procedures. Regardless of that, he is certainly sacrificing his family life and a comfortable bed in his obedience to orders. As he commences the recruit's training, he can be viewed as a disciplined link in the chain of command; one committed to obeying orders. These orders are designed to help recruits recognize the necessity of sacrificing personal comfort and giving up their civilian freedoms. Later in training, they will learn to sacrifice themselves for comrades if the need and situation requires.

---

[73] This quotation credited to J. Hudson Taylor is taken from a card picked up at a mission festival some years ago. No credit information is provided

Finally, orders are the sergeant's initial weapon and words the first bullets he fires at these recruits. He is the sergeant, a man in authority; when he speaks, you listen and heed without question. As noted in the opening chapter, military life is dissimilar to its civilian counterpart, as it does not tolerate either questions or demands for information. The sergeant is above questioning; disobedience to his command is unthinkable. His command: "Do as you are told" is an instruction concisely stated. It is obvious the first day of boot camp does not resemble or parallel in any way the situation enjoyed by similarly aged civilian students. When they head off to college, university or seminary, they receive a warm greeting from the president, a campus tour, introductory icebreaker games, and other welcoming perks of appreciation for having chosen the campus. Pampered students stand in sharp contrast to disciplined recruits.

As the initial euphoria of civilian campus life eventually stabilizes and the reality of study and routine are established, life in boot camp also adjusts to its own schedule. As readers will recall, discipleship characteristics were neither questionable nor optional; Christians must exert a similar regard for these five essentials. New Christians commence where the recruits began by gaining an understanding of Who is in charge; but Christians soon learn our Who always leads in love. The first necessity in any organized structure is to learn to serve under the leader's direction. The leader may hold a higher rank as in the military, serve as captain on a sports team, or carry *"the name that is above every name"* (Philippians 2:9), the name of Jesus. Paul writes concerning this, *"Everyone must submit himself to the governing authorities, for there is no authority except that which God has established"* (Romans 13:1). Leadership in government, institutions, companies, organizations, and religious establishments demands respect. Military training involves orders to be in physical training (PT) gear or combats (frequently seen camouflage uniforms) and in formation on the parade square by 0700 hours until dismissal at 1630 hours (i.e. 7:00 a.m. and 4:30 p.m.). This is only during basic training; maneuvers and deployments have a much more rigorous timetable that may or may not have a defined start or end time. Similarly, Christian life does not follow the typical 9-to-5 workday, and instead must be lived every waking hour. For readers to grasp the necessity of discipleship training, we look at these comments from Ogle. In this excerpt, he provides readers with a statement on the reality of training and a graphic picture of temporal warfare:

> In order to turn a civilian into a soldier, there must be rites of passage. They cannot only be physical in nature. First, not everyone is cut out to be a combat soldier. Some people simply do not have the physical or mental wherewithal to actually *fight* in a war…
>
> Every time a recruit fails, he strengthens the overall unit…A sense of pride fills the survivors. There is no sense of guilt, merely a feeling of accomplishment…they have passed where others have failed…an essential element of esprit de corps.[74]
>
> Second, Canadians must realize that combat soldiers are not called upon to be mere athletes. It is a trend in our society to liken athletic events to wars. Hockey teams "war" against one another. Boxers are referred to as "warriors." Football teams "do battle" on the gridiron. These terms are meant to add colour to sports and to glorify the participants, but really only belittle those who fight and die in the service of their countries. Sporting figures have rarely been killed while performing on the field. Young soldiers, however, perform on a different field. The battlefield is an arena that has no rules and no referee. The screams are not from cheering fans but from close friends who have been left mangled and smashed in the dirt. A brilliant tackle is not followed by a helping hand up. The downed man's head is smashed in with a rifle butt to make sure he never gets up again. The winning side does not hug the losing side. They are marched off to receive decent treatment or perhaps to be bayoneted on the roadside because they are bleeding too badly to walk any further. That is the reality of the soldier. He will not lose a game if he does not perform. He will lose his life.[75]

---

[74] Hawkins. 272. *esprit de corps*: loyalty and devotion uniting the members of a group
[75] Ogle. 39 (author's emphasis)

In the days that follow boot camp initiation, recruits are sifted to separate the wheat from the weeds, to draw on a biblical parallel (Matthew 13:24–30). In this case, the weeds are those unable to handle the mental and physical demands of military service—individuals some may view as weak.

Let's remember that individuals are diverse in their giftedness; no one is gifted for every career or every ministry. Here it becomes important to realize failure in military boot camp does not mark one as a loser; it simply means they would be better suited in a different career. The CF uses these failures to strengthen the remaining recruits by filling them with pride. It is an opportunity for those in command to enhance the self-image of successful recruits using the principle: "We (the strong) have done what others (the weak) could not do." Therefore, we are the champions! By contrast, Christian spiritual warriors hold to an opposing mentality; we depend on God for our strength. Paul writes about this weak/strong dichotomy:

> *But God chose the foolish things of the world to shame the wise; God chose the weak things of the world to shame the strong. He chose the lowly things of this world and the despised things—and the things that are not—to nullify the things that are, so that no one may boast before him.* (1 Corinthians 1:27–29)

> *But he said to me, "My grace is sufficient for you, for my power is made perfect in weakness." Therefore I will boast all the more gladly about my weaknesses, so that Christ's power may rest on me. That is why, for Christ's sake, I delight in weaknesses, in insults, in hardships, in persecutions, in difficulties. For when I am weak, then I am strong.* (2 Corinthians 12:9–10)

The structural difference is clear: military techniques build on self; Christian discipleship trusts in God. These contrasting plans demonstrate self-confidence versus faith in God. Nonetheless, parallelism is revealed by the motivation of each method as both strive for obedience to and confidence in leadership.

In the second part of the quote above, Ogle clearly depicts the horror of war. It is not some sports event for well-conditioned athletes; it is a life and death battle physically, mentally, emotionally, and even spiritually. Studies have proven that one human being does not, by nature, seek to take the life of another; even if he or she is viewed as a threat or an enemy.[76] This was observed in Afghanistan by Hope:

> The majority of soldiers when fired upon for the first time would seek to disengage back toward the "last safe place" they had occupied. After several encounters they repressed this urge but would be very reluctant to advance in contact (especially when separated from their [Light Armoured Vehicles]).[77]

The latter part of the passage from Ogle was included primarily because it describes more than temporal battles. It also graphically portrays spiritual warfare fought in a different type of arena where there are no rules, no referees; this is the reality of spiritual warfare. It, too, is not a game because the loss of eternal life is at stake for those not serving in God's tactical unit. As mentioned in a previous chapter, Jesus and Paul often turned to agriculture, military, and sports for illustrative purposes; however, both knew spiritual warfare, as with temporal battles, is not a sportsmanship game.

Ogle provides a peek inside boot camp in our military—insight only available to those who have experienced it. Basic training produces disciples of the CF, when a disciple is understood as someone who orients their entire lifestyle in obedience to someone in authority. As disciples, CF members are disciplined, committed, obedient, sacrificial, and knowledgeable in weapon handling. We may not normally think of discipleship as a component of military life or a subject on which the military can instruct Christians. I beg to differ! One objective of this

---

[76] The reference here is to the work of: Lt. Col. Dave Grossman. *On Killing: The Psychological Cost of Learning to Kill in War and Society*. New York: Black Bay Books/Little, Brown and Company, 1996.

[77] Lieutenant-Colonel Ian Hope. *Dancing With the Dushman: Command Imperatives for the Counter-Insurgency Fight in Afghanistan*. Kingston, ON: Canadian Defence Academy Press, 2008. 97

book is to assist Christian civilians to understand military life with an expectation that this knowledge will increase awareness of and support for our troops. My experience, as both a student and teacher, has revealed learning is not a one-way street running from teacher to student. So what you are about to read may sound shocking: as a Christian, this book offers you a chance to learn *about* the military from a non-military Christian; but it also provides you with an opportunity to learn *from* the military in regard to basic Christian discipleship characteristics. It is my belief that many Christians could profit from a period of boot camp training—Christian style, of course.

Let's be perfectly honest with one another about military boot camp and the Christian faith: both adhere to these five discipleship characteristics. They form the basics of boot camp and influence the thoughts of military personnel. Furthermore, they form the foundation of our faith and should influence our actions as Christians. Canadian troops as disciples of the CF are disciplined to recognize and respect authority; they believe in and are committed to the missions they undertake; they unquestioningly obey their CO; they are prepared if necessary to sacrifice their life for their comrades or in the interest of national security; and they know when, where and how to use a weapon. Similarly, Christians as disciples of Jesus Christ are called to recognize and respect the authority of the Holy Spirit, to commit to and fulfill our mission as spelled out in the Great Commission, to respond to the Holy Spirit's direction without question, make the supreme sacrifice if necessary, and know when, where, and how to use the Word, *"the sword of the Spirit"* (Ephesians 6:17). The two lists are parallel in thought but they contrast in terms of implementation. Nevertheless, CF disciples can teach and share with Christian disciples their techniques relating to discipleship. A question I pondered is, do the consequences of failed discipleship permeate Christian thought patterns? Should a Canadian soldier fail to comply with these requirements, they may endanger personal safety or the safety of comrades. Likewise in God's army, failure to perform Christian duties represents a danger to sisters and brothers in the faith.

The five discipleship characteristics applicable to both the military and Christian faith can be summarized and paralleled as follows:

Discipline in the military means being self-controlled and trusting in higher authorities. In Christianity, it means being self-controlled enough to read the Word of God and pray regularly. It also means yielding to the call of the Holy Spirit and His direction for our life.

Commitment in the military involves total faith in the mission and leadership of that mission. There is a similar parallel for Christians, as we are to faithfully follow Jesus Christ and be committed to His mission of sharing the Good News of salvation.

Obedience in the military requires a mind programmed to obey orders from the CO without question. The parallel in our faith is obedience to the commandments of Jesus Christ, our CO. His commandments are found in Scripture and obeyed through the power of the Holy Spirit.

Sacrifice for our troops may necessitate laying down one's life for comrades or more probably for the security of Canada. In the same manner, the Christian must be prepared to sacrifice everything, even life if necessary, for the cause of the Gospel. Paul is one example of what this may involve, as he writes:

> *I have worked much harder, been in prison more frequently, been flogged more severely, and been exposed to death again and again. Five times I received from the Jews the forty lashes minus one. Three times I was beaten with rods, once I was stoned, three times I was ship wrecked, I spent a night and a day in the open sea, I have been constantly on the move. I have been in danger from rivers, in danger from bandits, in danger from my own countrymen, in danger from Gentiles; in danger in the city, in danger in the country, in danger at sea; and in danger from false brothers. I have labored and toiled and have often gone without sleep; I have known hunger and thirst and have often gone without food; I have been cold and naked. Besides everything else, I face daily the pressure of my concern for all the churches"* (2 Corinthians 11:23–28).

Furthermore, Paul's experiences sound like what Canadian soldiers too frequently must endure. Should *"a good soldier of Christ Jesus"* (2

Timothy 2:3) anticipate sacrificing anything less? If this sounds a little extreme, Paul reminds us, *"in view of God's mercy, to offer [our] bodies as living sacrifices, holy and pleasing to God—this is [our] spiritual act of worship"* (Romans 12:1). In the Old Testament, imperfect blood sacrifices were required but they were replaced by the perfect sacrifice of Jesus. Nonetheless, sacrifice remains a Christian requirement.

Weapons in the hands of our military personnel must be cleaned, maintained and handled according to regulations. Christians also have a weapon, *"the sword of the Spirit, which is the word of God"* (Ephesians 6:17). However, this weapon is not oiled, greased, or polished—rather it is designed for reading, meditating upon, and memorizing. Like any weapon, it must be handled according to regulations which require knowing when, where, and how to use and share the Word.

As a culture, the military community is unique within Canadian society; as such, its basic training reflects in a family's military and civilian community. For the Christian community, another unique culture in Canadian society, our basic training reflects in our obedience to the *"King of kings"* (Revelation 17:14) and through our activity on His behalf. As Christians, every action we undertake is on His behalf, as Paul points out: *"Whatever you do, work at it with all your heart, as working for the Lord, not for men, since you know that you will receive an inheritance from the Lord as a reward. It is the Lord Christ you are serving"* (Colossians 3:23–24). What I have suggested is Christians can learn from military training, which can be demonstrated with one final quote from Ogle. He maintains it is incumbent on civilians to understand individual soldiers are:

> ...no longer just small cogs in a wheel, but men [and women] who could make the difference between victory and defeat.
>
> Generations of experience have taught NCOs [Non-Commissioned Officers] to recognize young men and women who simply do not belong in the military. They have also been taught to manipulate their charges either to remove these weak links or to harden them.
>
> Is this cruel? Yes. Illegal? Often. Necessary? For a nation

training its men and women to go to war, it is absolutely necessary.[78]

What can we learn from this observation by Ogle? This quote partially parallels my understanding of the Christian life. I maintain it is incumbent on Christians to understand they are not just small cogs in a wheel, but men and women who can make a difference in God's tactical unit. Beyond that point, the parallelism breaks down because there are none who do not belong or must be manipulated, removed, or hardened. However, each one of us does require training to become the best disciple of Christ possible since we will be engaged in spiritual warfare. Therefore, we can adapt the final thought from Ogle. Is our training cruel? No. Illegal? Never. Necessary? For the church training its men and women to go out and engage in spiritual warfare, training is absolutely necessary.

In closing this portion of the lesson, it must be noted our Canadian military has successfully trained and defended Canada for centuries. Having read about discipleship parallels between temporal and spiritual warriors, Christians should now be willing to investigate what troops can teach Christians. As CF disciples, their training and knowledge of discipline, commitment, obedience, sacrifice, and weapon handling could be beneficial to Christian civilian communities. If nothing else, they are role models to be imitated, although not always duplicated. However, before wrapping up our study of the initial training a temporal or spiritual trooper should consider, we need to further explore biblical parallels.

Biblical Parallels

There are numerous examples in the Old Testament of men and women who had some form of boot camp experience in their life. The following familiar individuals will be used for illustrative purposes: Jacob, Moses, Ruth, Samuel, David, Esther, and Daniel. Admittedly, they are not Christian disciples, but Old Testament examples of members of God's tactical unit. These members received training in the

---

[78] Ogle. 20

discipleship characteristics being studied. Furthermore, although they are foundational to the Christian faith, we can only imagine what their first day of boot camp may have been like. It is not the intent of this book to imaginatively attempt to describe their experiences. Below, I offer a concise overview on part of each individual's training experience.

Jacob. Boot camp normally separates families, placing recruits a considerable distance from their biological family. This was one of Jacob's experiences following his deceitful acquisition of Isaac's blessing intended for Esau, his first-born. Acting on his mother's advice, he fled to Haran to stay with her brother Laban (Genesis 27:43). In today's world, we may simply think that it's no big deal that Jacob was living in his uncle's household. However, when we consider the physical distance between Rebekah and Laban and lack of communication networks we enjoy today, this move becomes a very big deal. We also know his boot camp training covered a twenty-year period (Genesis 31:38, 41); during this time, he worked the first fourteen years to earn Laban's daughters Leah and Rachel as his wives. He disciplined himself to manage his father-in-law's flocks of sheep; he was committed to the task; he sacrificed his wages to earn his wives; and he skillfully handled the shepherd's rod and staff as well as his words to deceive his deceitful father-in-law. At this point one could digress and imagine what impact this had on managing his own flock of two wives, two concubines, twelve sons, and one daughter who is mentioned; but that would take us off topic.

Moving on to Moses, he experienced family separation as well, but in a different style of boot camp than Jacob. As an infant he was adopted by Pharaoh's daughter, lived in Pharaoh's palace, and was trained as a possible heir to the Egyptian throne. All this occurred because his parents wanted to protect him from Pharaoh's order to drown Hebrew males in the Nile River. His mother ingeniously set him afloat upon that river. It resulted in one of two boot camp experiences. The other came at forty years of age (Acts 7:23–24) when he fled to Midian after he murdered an Egyptian (Exodus 2:11–12). At Midian, Moses served a forty-year term in a boot camp comparable to the one Jacob attended (Acts 7:30). The outcome of the two boot camp experiences was the preparation of Moses to confront Pharaoh and eventually deliver the descendants of Abraham from Egyptian slavery. To accomplish this feat, Moses had to be committed and obedient to God and His mission. He made great personal sacrifices and handled God's word skillfully to complete the task when dealing with Pharaoh and later the Israelites in the wilderness.

The third individual, Ruth, does not have a documented boot camp experience. As a result, we must work with some assumptions. Many biblical commentators assume that Elimelech's family remained in Moab *"about ten years"* (Ruth 1:4). They further assume, since his sons died without leaving children, their marriages must have been near the end of that period. By accepting these assumptions, one can concluded that Orpah and Ruth, two Moabite women, had little time to develop an understanding of the Jewish religion of their husbands. In addition, the Moabite god was Chemosh, who accepted human sacrifice—a practice detestable to the God of the Jews. After her husband died, Ruth made a decision to trust Naomi's God, the God of her mother-in-law. Today, knowing the full story, which includes Ruth's placement in the genealogy of Christ, we recognize that she was obedient to a Higher Power. She may not have fully understood Him, but she obeyed Him without question. We also find her making a self-sacrificing commitment to God and Naomi in this declaration: *"Don't urge me to leave you or turn back from you. Where you go I will go, and where you stay I will stay. Your people will be my people and your God my God. Where you die I will die, and there I will be buried. May the Lord deal with me, be it ever so severely, if anything but death separates you and me"* (Ruth 1:16–17). These words are the weapon that stopped Naomi from continuing to urge Ruth to return to her people and her gods as her sister-in-law had just done. Is it possible Ruth experienced a form of private personal training directly from the God she did not fully know? It appears this may have been the case, because she demonstrates all five of the discipleship characteristics we are examining. The book of Ruth is well worth spending some time studying.

Next, we examine Samuel's experience. Like each of these biblical personalities, he received unique boot camp training. He was the fulfillment of a prayer by his mother, Hannah. After he was weaned, and as she had promised God, she gave him to the Lord for his entire life. The young *"boy ministered before the Lord under Eli the priest"* (1 Samuel 2:11). What is unique in Samuel's training is his teacher, Eli, taught him well but could not do likewise with his out-of-control sons. While Samuel learned discipleship qualities, Eli's two sons grew increasingly disobedient, rebellious, and wicked in their relationship with God and Eli. This prompted an announcement from a man of God claiming the death of Eli's sons would occur on the same day (1 Samuel 2:27–34). We read of their death and other events in 1 Samuel 4:12–22. Samuel stands in contrast to Eli's sons in all five characteristics.

Samuel had the privilege of anointing our next person of note to be king following the death of Saul, Israel's first king. Samuel visited the home of Jesse and overlooked the seven older boys in favour of the youngest son, David. He was brought from the pasture fields to be anointed (1 Samuel 16). In these fields, David had been tending his father's sheep, providing David with shepherding experience parallel to Jacob and Moses. It was not long until he found himself in Saul's service; here we read of his many accomplishments. In terms of military achievements, David is best remembered for the slaying of Goliath. Once again, David's training is not documented, but his life story speaks of commitment to the flock, obedience, family separation, and sacrifice. In addition, we have his skillful use of words found in many of the Psalms. As his life unfolds in the pages of Scripture, David is revealed as a man constantly in conflict. More significantly, Paul describes him this way while preaching in Antioch: *"After removing Saul, [God] made David their king. He testified concerning him: 'I have found David son of Jesse a man after my own heart; he will do everything I want him to do'"* (Acts 13:22). David truly is a man who successfully graduated from God's boot camp.

Like Ruth, Esther makes an interesting study, but again one must read between the lines to identify her boot camp experience. Her training is obvious in the fourth chapter as she agrees to sacrifice her own safety for the Jewish people with the statement: *"And if I perish, I perish"* (Esther 4:16). Orphaned as a child, Esther was raised like a daughter by her cousin Mordecai (Esther 2:7). Through an unusual series of events, she replaced Queen Vashti; but a plot soon developed designed to destroy the Jewish people. Esther had not revealed her Jewish heritage when she became queen; thus, one might view her as uniquely positioned to foil the plot. However even as queen, she would be subject to death if her heritage was discovered. She also risked death if she approached the king uninvited, and was thus caught between a rock and a hard place. However, it was because she was prepared to defy the king's order and approach him that she made the above statement. Her action on behalf of her people was a decision only she could make, yet she sought counsel from Mordecai. He was her cousin and trustworthy because he was the one who had committed to and sacrificed for her as a child. She could not disregard his instruction and their relationship proved to be a two-way commitment; for Training Sergeant Mordecai, she was prepared to make the ultimate sacrifice. If one takes time to read her dialogue with King Xerxes and Haman, it becomes apparent that Esther was skilled with her weapon of words (Esther 5, 6 & 7). Esther's actions exemplify her training, which may not be categorized but is evidenced through her activity.

Finally, we arrive at Daniel. In the first chapter of the book of Daniel, it is revealed he was part of the royal family of Israel and he possessed a list of impressive qualities. In an order from Nebuchadnezzar, King of Babylon, Daniel and other members of the royal family were conscripted for training in the king's service. His first training was in Israel. As a member of the royal family, he would have been through a type of royal boot camp. Now in addition, Nebuchadnezzar was demanding further instruction. One requirement for the participants was to partake of the Babylonian royal food and wine—a condition which would defile Daniel before his God. As a result, Daniel and three other members of the royal family devised a strategy of compromise and struck a deal with their guard. Their ten-day trial period proved successful and these men completed three years of training undefiled in the sight of God. These men were not trained for military duty but rather in the language and literature of Babylon; a different style of boot camp, but one in which they remained loyal to God, obeyed an earthly king, and fulfilled the characteristics discussed in this chapter.

There are fewer New Testament individuals to draw upon, because less is known about them; although we still find diversity in boot camp teachers, techniques, and locations utilized. The largest portion of the New Testament is given to us by Paul through his letters; we can add to those letters Luke's account of Paul's life as found in Acts. Unlike the Old Testament, the New Testament does not portray the history of the Israelites over centuries; rather, it provides insight into the first century of development of the Christian church. This means the two Testaments represent a contrast between sharing a detailed family history and foundational historic information on the seed of God's church. At the same time, both parallel one another in describing the ongoing activities of God in our world. Therefore, to continue this review in the New Testament we will begin with Jesus and the original disciples, Paul and John Mark.

Following a visit by Jesus to the temple at twelve years of age, Luke writes, *"And Jesus grew in wisdom and stature, and in favor with God and men"* (Luke 2:52). We all commence our training at birth; Jesus was no exception. Furthermore, as with many even to this day, he received religious instruction. This type of education does not replace the need for God's boot camp experience; what it does is influence one's life in a manner parallel to boot camp. In Scripture, we find that idea expressed as a general principle in this way: *"Train a child in the way he should go, and when he is old he will not turn from it"* (Proverbs 22:6). My view is that boot camp for Jesus began immediately after His baptism with the forty-day temptation in the desert by the Devil. This was the pinnacle of His spiritual warfare boot camp. During those forty days, Jesus received training which: prepared Him to maintain a disciplined life of prayer and ministry (Mark 1:35–39); fortified His commitment to the mission, allowing Him to pass the Gethsemane test (Matthew 26:36–46); prepared Him to be obedient to *"even death on a cross"* (Philippians 2:8); strengthened Him to make the journey to the Cross, as He *"resolutely set out for Jerusalem"* (Luke 9:51); and proved the power of His only weapon—the Word of God (Luke 4:1–13) as He responded to the Devil's temptations. Jesus was tested in all five discipleship characteristics we have been discussing in this chapter and He graduated with a perfect grade point average.

What I find encouraging about the original disciples is they spent three years in training under Jesus and still did not get it all figured out. It was not until the Holy Spirit was poured out that things began to make sense to them. Obviously, there is still hope for the rest of humanity. We do see one interesting technique used by Jesus which is often applied today in many training situations, including our military. Before sending out the Twelve (Luke 9), Jesus demonstrated miracles and healings (Luke 7:11–17); next He allowed them to participate, as in the calming of the storm (Luke 8:22–25); finally, He released them with power and authority to drive out demons, cure diseases, preach the kingdom of God and heal the sick (Luke 9:1–2). This demonstrates the common three-step principle: first, I will do it for you; second, you will do it with me; and finally, you will do it while I supervise.

Paul suffered greatly for his faith in Christ, as recorded in the earlier summary on sacrifice. So where did he go to boot camp? Late in his ministry, while in Jerusalem, he said, *"I am a Jew, born in Tarsus of Cilicia, but brought up in this city. Under Gamaliel I was thoroughly trained in the law of our fathers and was just as zealous for God as any of you are today"* (Acts 22:3). Gamaliel was one of the most respected teachers of his day, so Paul was well versed in Hebrew Scripture, our Old Testament. After his conversion, Paul *"spent several days with the disciples in Damascus"* (Acts 9:19) but due to a conspiracy to kill him, he left for Jerusalem. Shortly after his arrival, a second attempt was made on his life and he went to Tarsus. According to his testimony, he *"went immediately into Arabia and later returned to Damascus. Then after three years, [he] went up to Jerusalem to get acquainted with Peter and stayed with him fifteen days. [He] saw none of the other apostles—only James, the Lord's brother"* (Galatians 1:17–19). Commentators generally agree Paul spent most of his three years in Arabia, away from Damascus. This was probably his boot camp—a desert encounter with the full power of the Holy Spirit. We know from Luke's account of his life in Acts and Paul's letters that he practiced the five discipleship traits we have been exploring.

Finally, we look at John Mark, author of the New Testament book of Mark. This is a man with much to tell each of us today. In Peter's first letter, dated in the early 60s, he closes with, *"She who is in Babylon, chosen together with you, sends you her greetings, and so does my son Mark"* (1 Peter 5:13). Commentaries point out that Peter's affection for this individual prompted him to call Mark "my son," though there was no evidence of a kinship. In addition, the book of Mark is thought by many to be an interpretation of Peter's sermons. He also travelled with Paul and Barnabas as their helper on their first missionary trip (Acts 13:5). We may not know much about his boot camp experience, but he had some of the best instructors available. We also know early during that first missionary trip he *"left them to return to Jerusalem"* (Acts 13:13). Furthermore, before a second trip, Paul and Barnabas had a disagreement over taking John Mark with them. Paul interpreted Mark's premature return to Jerusalem as a clear case of deserting the mission or, in military terms, being absent without leave (AWOL). As a result, Paul and Barnabas parted company with Barnabas, taking John Mark and Paul heading out with Silas on separate missionary trips. Later in his ministry, about 60 A.D., Paul writing to the Colossians sends greetings to that church from, among others, *"Mark, the cousin of Barnabas"* (Colossians 4:10). About 66 A.D., Paul writing a second letter to Timothy instructs him to: *"Get Mark and bring him with you, because he is helpful to me in my ministry"* (2 Timothy 4:11). Whatever problems Paul had with Mark earlier, they apparently had been resolved and Mark was now an asset to Paul. So, did Mark desert the first mission? Or, was he not fully and properly trained for the mission field? Was Paul too quick to pass judgment on this man? For these and other questions, one could develop hypothetical answers but the truth is we do not—and never will—know what occurred between these two men. The reason for including Mark is to demonstrate a point: a boot camp failure, such as the one Mark apparently had, is not the end of the story with God. Failure in military boot camp may terminate a career, but in God's tactical unit there are none who do not belong. Yahweh is the God of second chances; in His tactical unit, there is always hope for a bright future.

This list of military boot camp parallels was not selected to demonstrate exact parallels with the military example used to open the chapter—obviously, it cannot and does not do that. In the military, it may be essential for every recruit to receive an identically structured boot camp program. However as shown above, God instructs each of us through a personally tailored format, at various stages of life, in different locations and with diverse techniques. God trained these individuals of the Bible in sheep's grazing meadows, the palaces of foreign rulers, religious institutions, an adversary's military, and by divine intervention in people's lives. For God's tactical unit, His training has certain minimal requirements but maximum standards.

As we prepare to continue with our lifestyle tour, take time to reflect on living with war or its threat. The poem opening chapter four begins with a number of different paraphrased statements from individuals who were interviewed and mailed letters or sent emails to Canadian authors. The poem concludes with a paraphrase of Matthew 24:6–14, a passage validating the poem's title.

War Declared!
FOUR

WAR IS INEVITABLE
Living with war, or its threat, is a reality in military life today.
So how does one function to keep this worry and stress at bay?
First, from the books listed at the back; some folks have this to say:[79]
A military wife understands and appreciates little things in each day.
When front line news it does arrive and fear seizes every heart;
Be it good or bad, all will pray their loved one did not depart.
Sometimes the horror is to survive when other comrades are lost;
If it impacts your friends, you suffer with those bearing the cost.
Survivor guilt this is called, a front line troop reality—no doubt;
It leads to Post-Traumatic Stress Disorder with problems all about.
Troops deploy, say good-bye, and the home front shares the terror.
But this is war; we're all in one boat, so show you are an enduror.
War affects each differently, be it hubby, dad, mom, wife, sis or son.
Some say: I did enough! Or ask: Why me?—What have I done?
After training for so many days, duty answered makes one glad.
Wishing they could be there makes partners both proud and mad.
One may say, "It's their job!" For what, your safety to get killed?
Truth is all suffer loss; both ours and enemy blood will be spilled.
Protestors insist: End the war! Bring them home! It's best for all!
On the other side: Your killing morale! While they give their all!
The debate goes on with protests proclaimed from every side,
And yet we know God upon His throne high above does abide.

---

[79] The following are paraphrase comments from books in the Bibliography

God sent his only Son; an act of His great Love.
The One delivering the Word to us from above.
Second, what did He say of the future to come?
Of war and threats of it be not startled some.[80]
For all this will happen before the final hour;
Nation against nation; power against power.
Famines and earthquakes in different places;
Beginning of birth pains, these are the traces.
Persecuted and even put to death you will be;
Hated by all nations, this because you love Me.
Many from the faith, they will certainly turn;
Betraying and hating each other they'll learn.
Of false prophets, great numbers will appear;
Deceiving people with words to tickle their ear.
With increased wickedness, love will grow cold.
But for he who stands firm to the end and is bold,
He is the one to whom salvation will surely come.
My Gospel must be preached to all, not just some.
My testimony is for nations the whole worldwide;
Then the end it will come placing you at My side.

WAR DECLARED!

In recent years, most Canadians may not have read a newspaper headline like the title of this chapter. I do not recall such a headline in Canada regarding the war on terrorism. Rather, we did what was necessary without a great deal of fanfare. Now how typically Canadian is that! Regardless, in the previous chapter Ogle claimed: "It is a trend in our society to liken athletic events to wars."[81] There is also another trend in our society employing the word 'war' for dramatic effect. Let me be clear that the war on terrorism is a genuine theatre of war—one that has cost many Canadian soldiers their life. On the other hand, when government leaders, other civic authorities, and newspapers announce a war on crime, drugs, homelessness, poverty, unemployment, etc., they do this

---

[80] Paraphrase of Matthew 24:6–14
[81] Ogle. 39

solely for theatrics. These are not wars but rather an increased focus by a certain department on a current social problem. To paraphrase Ogle, the word 'war' is meant to add colour to these actions; in reality, it only belittles those who fight and die in war and in service to our country. Thus, this chapter will focus on living with war; real war, as with the declaration Canada made on September 10, 1939 against Germany, as well as the more recent Middle East conflicts. In addition, it will consider the ongoing stress of living with the threat of war; both of these are continually part of military life. The main objective here is to attempt to acquire some understanding on the effects war has in our military families' lives.

Apart from our veterans and military families, few in Canada today can relate to the issue of living with war or its effects. As a member of the Canadian boomer generation, I have not experienced compulsory military service or conscription. In the United States, it was very different with many my age serving a mandatory military term, primarily a result of the war in Vietnam. Further compounding efforts to understand how one copes and lives with this conflict is our natural tendency not to discuss unsettling circumstances. Through my ministry, I have had the pleasure of working with a number of military families and one thing that has impressed me is how they can accept living with war or the threat of war as part of their job description. Truth is, often it is better to allow "what if" questions to remain dormant or say as some do: "I will cross that bridge if and when I get to it." In other words, it may be impossible for civilians to obtain insight into the effect war has on military families by dialoguing with them. That does not, however, eliminate the need for civilians to be aware of this stress factor in the lives of military personnel and those who are associated with them. In fact, I think a marvelous opportunity exists in this field for a professional sociologist to conduct a study and report the findings. The information could prove a valuable tool for raising support for our military.

In the absence of sociological research, an alternative is to scan comments from various Canadian publications. After reading the books listed in the Bibliography, I realized that the personal stories shared are honest and open about individual experiences and feelings. Primarily, these involved past wars or peacekeeping duties. Thus, I offer this sweeping overview from my reading. The full spectrum of emotions was observed, from paranoia at one end, through panic, to paralysis, into nonchalance and finally total resolution to see the mission fulfilled. Some detailed only one emotion as if they were stuck there. Most people, however, reported progressing through a range of emotions, starting and ending in different emotional states. In society today, twenty-first century families can expect to experience their own emotional reactions with deployments to Afghanistan. Personal responses may have similarities to those detailed by each author; nonetheless, they will be uniquely individual. This reality reveals a truth readers must be aware of: events in life affect each of us differently! War declared or threatened will raise an array of emotions, and prompt a variety of coping mechanisms. For those left on the home front, the emotions of anxiety and fear cannot be ignored as one's spouse is deployed to a war zone. It is important for the Christian civilian community to support our troops and their families at these crisis points in their lives. We are not called to judge the conflict or individual emotional reactions to circumstances by military family members. Nonetheless, Christians must be prepared for all extremes and every point in-between. In 2004, Collier wrote: "After chatting with many ladies whose partners were involved in the first tour in Afghanistan, I realized that our fears for our partners' safety had increased 100 percent."[82] Recently, our nation wrapped up the engagement in Afghanistan. The fears of our military families today may well have multiplied exponentially, considering the massive loss of Canadian lives in that conflict. As military families live with war or with the threat of war, they need our support and prayers as did the veterans who went before them. In addition, may we never forget our veterans still deserve our prayers and support! The majority of our troops may have recently been brought home; however, living with war remains a reality. It is a certainty that in the not-too-distant future Canada will once again be engaged in a world conflict or a peacekeeping tour. As readers may recall, past tours of this nature were often not very peaceful for our troops.

---

[82] Dianne Collier. *My Love, My Life: An inside look at the lives of those who love and support our military men and women*. Carp, ON: Creative Bound Inc, 2004. 85

In addition to my reading on the impact that war (or the threat of war) places on military life, I have had the privilege of interacting with a family from Rwanda. This opportunity arose as the result of a generous congregation pledging support for one year to the family. It has been my privilege to interact with this congregation over the years, but even more so during their yearlong sponsorship. Through meeting and talking with the family, I can only conclude they have lived for years with many of the issues discussed in this chapter. There is parallelism between their plight and military life, although I acknowledge there is a difference in settings and circumstances. As refugees, their life held an uncertainty of what tomorrow may bring and it accompanied them as a daily companion. The anxiety of wondering who would or would not return home at day's end created dark shadows on the wall of their home. A constant threat hanging over their head was the fear and turmoil of being uprooted and fleeing the regime of terror ruling their nation. Finally, they endured the discomfort of refugee life in a strange and unfamiliar land; this was the family's greatest sorrow. The family lived with war, its threat and its consequences for years, waiting for an opportunity for a sponsorship to Canada. Does this not reflect a soldier and his family's lot as well? One of living with an uncertain tomorrow, wondering if a loved one will return from active duty, being uprooted due to war and experiencing the discomfort of living in a strange and unfamiliar land. Admittedly, these are not exact parallels but they do have application in helping civilians realize the life military families endure for our freedom and security. My Christian brother and his family transitioned to a new life and possess a peace evident in their countenance—peace which passes all understanding. My experience has been that our military families exhibit this same sense of peace as they live daily with the threat of war or its inevitable reality. Surely their peace connects to this quote on the cenotaph in my home town: "That which we do for ourselves alone dies with us: that which we do for others lives on and is immortal." Our troops sacrifice a great deal for others! However, one thing is certain: the battlefields of today differ greatly from the trench warfare of the Second World War. High-technology vehicles, computer-directed bombs, and specially trained warriors reduce the feasibility of conscripting the general public into military service. Fortunately, we are blessed in this nation with men and women who have responded to the call to serve. They are willing to do so and often make the supreme sacrifice of their life both in war and peacekeeping. They are provided with and capable of utilizing the latest computer technology and are trained specialists in modern warfare. Thankfully, these competent professionals engage in this sophisticated warfare in order to ensure Canada's freedom!

There is yet another dichotomy for Christians to reason through regarding war or its threat. In Scripture, Christians are called to hate sin and love the sinner—a difficult task to carry out. The military is a reflection of this concept. It may be just as difficult to hate war and love the troops, but there is an obligation on our part to do so. After all, we will always have sin and sinners, as well as war and troops. To verify this, Jesus spoke of war as a sign of the end of the age; an indication all four—sin, sinners, war and troops—will be with us until He returns:

> *You will hear of wars and rumors of wars, but see to it that you are not alarmed. Such things must happen, but the end is still to come. Nation will rise up against nation, and kingdom against kingdom. There will be famines and earthquakes in various places. All these are the beginning of birth pains.*
>
> *Then you will be handed over to be persecuted and put to death, and you will be hated by all nations because of me. At that time many will turn away from the faith and will betray and hate each other, and many false prophets will appear and deceive many people. Because of the increase of wickedness, the love of most will grow cold, but he who stands firm to the end will be saved.* (Matthew 24:6–13)

Jesus secured believers peace with God; He did not come to bring a literal peace on earth among all people. He reminds his followers, *"A man's enemies will be the members of his own household"* (Matthew 10:36). Thus, to elaborate on the fact that you can hate war but love the troops, it is as legitimate to protest war as it is logical to object to sin, because war is the result of our sinful nature. We can hate sin and war but both will be with us until Christ's glorious return brings the defeat of those who oppose Him. At the same time, Christians have an obligation to love sinners and our troops. Love means supporting our troops as they live with war or the threat of war while defending our freedoms; in particular our freedom of speech and freedom of religion. I want to reiterate, Canada has withdrawn from Afghanistan, but Afghanistan is sure to be eventually replaced by another hot spot somewhere…meaning the threat of war will continue to loom over the heads of our military personnel and their families.

Now, as usual, to close the chapter I draw on biblical parallels to demonstrate the emotional upheaval caused by living with war or its threat.

Biblical Parallels

One of the earliest and most descriptive biblical parallels of living with war or its threat is delivered by *"a prostitute named Rahab"* (Joshua 2:1). Joshua sent two men as spies into Jericho; it was part of his plan for strategizing an attack on the city. These spies were discovered and forced to take refuge in Rahab's home. She described to them what it was like for her people to live with an invading nation at their doorstep:

> *"I know that the Lord has given this land to you and that a great fear of you has fallen on us, so that all who live in this country are melting in fear because of you. We have heard how the Lord dried up the water of the Red Sea for you when you came out of Egypt, and what you did to Sihon and Og, the two kings of the Amorites east of the Jordan, whom you completely destroyed. When we heard of it, our hearts melted and everyone's courage failed because of you, for the Lord your God is God in heaven above and on the earth below. Now then, please swear to me by the Lord that you will show kindness to my family, because I have shown kindness to you. Give me a sure sign…"* (Joshua 2:9–12)

Today we know Rahab and her family were spared in the destruction of Jericho because she hid the spies. We also know she appears in the family tree of our Lord and Saviour. Where the passage parallels this chapter is in its description of the emotional reaction of the citizens to an invading army—they melted in fear, their hearts melted and their courage failed. Undoubtedly, Jericho was a highly emotionally charged city.

It was not always those under threat by Israel who melted with fear and lost their courage. In the days of Saul and David, the Philistines constantly waged war with Israel and on one occasion the giant Goliath confronted and tormented them to the point where *"Saul and all the Israelites were dismayed and terrified"* (1 Samuel 17:11). *"When the Israelites saw the man [Goliath], they all ran from him in great fear"* (1 Samuel 17:24). The tables were turned and now the army of Israel was dismayed, terrified, and in great fear. Thankfully, by God's power a young shepherd named David defeated this giant that stood *"over nine feet tall"* (1 Samuel 17:4). This same David in time became King of Israel and wrote some of the most beloved Psalms; yet not all were Psalms of praise. A number can be classified as laments, which reveal negative emotional states. However, David always concludes them by reflecting on God's goodness and declaring his trust in Him. For instance, in fear this mighty warrior cried out to God in Psalm 55:

> *Listen to my prayer, O God, do not ignore my plea; hear me and answer me. My thoughts trouble me and I am distraught at the voice of the enemy, at the stares of the wicked; for they bring down suffering upon me and revile me in their anger. My heart is in anguish within me; the terrors of death assail me. Fear and trembling have beset me; horror has overwhelmed me.* (1–5)

David is facing an enemy and his mind is troubled, he is distraught, his heart is in anguish, terror attacks him, fear and trembling over take him, and horror overwhelms him. King David, himself, vividly expresses the experience and emotions of living with war or its threat. The words of Rahab and this Psalm by David echo in the comments shared by people today. Possibly no sociological study is needed. We can discover the emotional impact of war or its threat from Scripture, in the social range from prostitute to king. If we examine other readings, we will find throughout history people have shared similar feelings about war or its threat.

As we search the New Testament for parallels, they become more difficult to find. Consider this question: was Jesus called to deal with the effects of living with war? I would reply, most definitely! Jesus was enmeshed in spiritual warfare to a degree we are unlikely to ever encounter. His family endured refugee status to escape the war raged by the furious tyrant King Herod against boys two years old and under (Matthew 2:16). We can add to this His desert struggle with the Devil (Matthew 4:1–11). However, nowhere do His emotions show more than in His Garden of Gethsemane prayer:

> *He took Peter, James and John along with him, and he began to be deeply distressed and troubled. "My soul is overwhelmed with sorrow to the point of death," he said to them. "Stay here and keep watch."*
>
> *Going a little farther, he fell to the ground and prayed that if possible the hour might pass from him. "Abba, Father," he said, "everything is possible for you. Take this cup from me. Yet not what I will, but what you will." (Mark 14:33–36)*

Three times Jesus prayed this prayer. He was distressed, troubled, and overwhelmed with sorrow as he prayed to His Father before entering into the battle in which He sacrificed His life for each one of us. Jesus was fully God and fully human and He functioned in one accord with the Holy Spirit—a state of being which may prompt questions about His distress, troubled spirit and sorrow. What we must remember is the "cup" of which he spoke contained all the sin of humanity throughout history to that point plus all the sin of humanity from that point through to the end of this world as we know it. In other words, it contained the sin of every human being who has and ever will live on earth. A certain cough syrup-maker advertises that when you take a spoonful of their product you will discover it tastes bad but it really works. Jesus knew the cup He was about to drink down to its last dregs would make that cough syrup taste as sweet as honey by comparison. Praise God, Jesus took that cup and by doing so it really worked to bring about our forgiveness before a Holy God who cannot look upon sin. However, the sacrifice is only effective if we accept what Jesus did on our behalf on the cross, turn from our sinful ways and strive to follow Him in total obedience, which will result in us sharing in His cup.

The point about Christ being filled with the Holy Spirit and yet feeling the effects of facing the greatest battle in all history has been raised for a reason. Through the Holy Spirit, believers have the power to overcome the emotions Jesus expressed; but as already stated, Jesus was called to endure far beyond human abilities. He was more than justified in praying as He did. Thus, when we look at the accounts of the disciples and Paul, it is difficult to find emotional reactions to the threat of war. Again, we must remember their battles were spiritual in nature, not temporal, in contrast to conflicts mentioned from the Old Testament. In Luke and the other Gospel accounts, we read about Peter denying Jesus three times, and then how *"he went outside and wept bitterly"* (Luke 22:62). It was a cowardly act for him to run from or abandon his Lord. Even though Jesus had warned him this would happen, in reality Peter chose to go AWOL on the front line. Before we find fault with Peter, however, we had better examine our own life for parallels to this incident. Making excuses or blaming others is part of our human nature. This trait goes back to Genesis when Adam blamed God and Eve, Eve blamed the snake, and the snake didn't have a leg to stand on. If we are honest with ourselves, when caught violating a law of the land, we tend to find someone or something to blame; that is to say, we make an excuse for our human behaviour. We thus compound our failures with excuses or by blaming others when we should be accepting responsibility and seeking forgiveness. Peter knew he had failed Christ and went outside to weep. His denial of Christ proved he was human; his weeping demonstrated acceptance of responsibility and seeking of forgiveness. As a result of Peter's repentance, our risen Saviour reinstated Peter (John 21:15–19).

Once Christ ascended and the Holy Spirit was poured out, the demeanour of the disciples changed. Similarly, we see this in Paul after he encountered the full power of the risen Christ. My personal opinion is that these men continued to experience fear, anxiety, terror, melting hearts, and loss of courage, but chose not to emphasize these as they wrote. This was not a denial of their feelings and emotions, but rather a decision not to draw attention to what they overcame through the power of the Holy Spirit. Peter, the coward in the courtyard, became a champion speaker before a massive crowd in Acts 2:14–41 and again before the Sanhedrin in Acts 4:8–21. Paul, at that time known as Saul, approved the stoning of Stephen and began to destroy the church (Acts 8:1–3). After his encounter with Christ, however, Paul set out on a life of missionary work facing all the hardships mentioned earlier, apparently with little fear. Some may ask, did Paul ever encounter fear and troubled feelings? I believe the answer to that question is yes! Consider the following passage as Paul finds himself in the midst of a violent storm that threatened to sink the ship carrying him to Rome. He tells the sailors:

> *"Men, you should have taken my advice not to sail from Crete; then you would have spared yourselves this damage and loss. But now I urge you to keep up your courage, because not one of you will be lost; only the ship will be destroyed. Last night an angel of the God whose I am and whom I serve stood beside me and said, 'Do not be afraid, Paul. You must stand trial before Caesar; and God has graciously given you the lives of all who sail with you.' So keep up your courage, men, for I have faith in God that it will happen just as he told me."* (Acts 27:21–25)

The angel of God appeared to Paul during the night apparently for no reason other than to settle Paul's fear, since this was the only issue the angel addressed. What Paul was specifically afraid of is unknown, but was significant enough to prompt God to comfort him. As with Peter and the others, he chose not to be overtaken or controlled by his emotions. It may be difficult to locate emotions of living with war in the New Testament, but these men were as human as anyone living today. They had emotional reactions to circumstances just as we all have and will continue to have. However, with the love of Christ we can help one another through these crises.

From the stress of living with war or its threat, we will study another stress creator in the form of the moving experience of being posted. The poem introducing chapter five documents the wanderings of Abraham and his family.

Posted—Again!
FIVE

Did Abraham Get Posted? [83]

With Dad set out from Ur, taking wife and nephew too.
Stopping in Haran, till Dad's death notice it came thru.
At seventy-five his call from God was heard and said:
'Your country, people and household prepare to shed.
There's a land I'll show you, it's there you will dwell.'
To Shechem's tree building to the Lord an alter swell.
It wasn't long until east of Bethel he quickly did race,
Pitched a tent, built to the Lord an altar in open space.
Off to Negev he then headed, being posted once again.
Tummy growls—famine here, he just couldn't remain.
To Egypt with wife and nephew, his postings did proceed;
Sarah claimed as a sister his life to secure. Oh such a deed!
The trick worked well it seemed but had its repercussions;
The Lord, He was not pleased by Pharaoh and his actions.
So up from Egypt, re-posted to Negev again for a while.
Here he added more livestock, silver and gold to his pile.
Prosperous, nephew too, t'was back to Bethel they trod.
Disputes arose but they fought not, in obedience to God.
'Lot, which way?' said Abraham, 'left or right it's up to you!'
'I'll take the fertile Jordan plain,' he said, 'the rest is for you!'
On new postings off they set; Lot took the pleasant, easy way.

---

[83] Summary of Abram's journey recorded from Genesis 11:31 to Genesis 25:11 (During Abram's life, God changed his name to Abraham and Sarai's to Sarah)

At the great Mamre trees of Hebron, Abraham decided to stay.
A home for a lengthy time and scene of significant events.
Sodom attacked, Lot's family seized by four kingly gents.
First army of God's people called up to save Lot and all;
Covenant, circumcision, a son promised his own to call.
Ishmael was born, grew up but he not the promised son;
His mother Sarah's maidservant, not God's chosen one.
Ishmael thirteen as Abraham and Sarah laugh at the plan.
Birth miracle foretold, Isaac, God's one to grow the clan.
Sodom and Gomorrah God ruined as judgment for sin;
To mountains Lot and daughters fled there to reside in.
For Abraham it's back to Negev for yet another posting,
Then to Gerar where his old Egypt trick again did ring.
So on to Beersheba for Isaac's birth and record the day,
But as for Ishmael, it was from here he did get sent away.
Beersheba was a posting of length but at an awesome price!
God commanded, "The son of promise offer as a sacrifice."
Foreshadowing God's only Son who on the cross He died;
For Abraham's obedience a ram in a thicket He did provide.
On to Kiriath Arba (Hebron), a posting of sorrow and joy.
Sarah's death and to relatives a servant Abraham did deploy.
The town of Nahor he did go bearing gifts for a bride so fair.
Rebekah his camels watered, a sign her and Isaac could pair.
On sharing with Rebekah's family of old Abraham's plan,
He's set to return, the bride-to-be anxious to meet her man.
This marriage Abraham had planned; an event of pure joy.
Abraham too, a new wife he took bearing him boy upon boy.
His final posting did arise at hundred and seventy-five years,
Sarah he'd join once again as the family fought back tears.
But the postings continued as son Isaac picked up the baton,
And grandson Jacob after him, and he did pass it on and on...

This parallels the way military postings go
From Gagetown to Petawawa to old Shilo.
To the west coast or Europe or even the States,

Never sure just what proper protocol dictates.
For training and maneuvers what might it be?
Borden, Meaford, or Wainwright, let me see!
But wait! Back up! For all got the same start,
In Recruit School, there we all gave our heart.
Or will it be in old Trenton town I'm found,
Waiting for a deployment flight am I bound?
And when we get settled in one cozy spot
Around the circuit once again we will trot.
Now this is just a short list of some stops,
But the one we're posted to just now is tops!
Although the gate I'll have to check to see
Which of the many CFBs this one might be.

POSTED—AGAIN!

Requisite relocation is part of life for families in our military. Generally in the past, this feature of military life was the number one stress factor, caused by moves every two or three years. It may have impacted the family even more than living with war or its threat. Today, postings remain a key stress component for military families, but are not the primary source of stress. The earlier quote by Collier points out the principle stress factor today is deployment. There is no doubt about this, as the next chapter will clearly demonstrate. In many ways, deployment is a combination of chapters four and five plus the addition of unknown after-effects like Post-Traumatic Stress Disorder (PTSD). PTSD merits its own chapter and is discussed under Returning Home.

There is a clear-cut distinction between a posting and deployment. Whereas a posting relocates the entire family to a different Canadian Forces base (CFB), deployment is the assignment of military personnel for duty, unaccompanied by their family. This duty or engagement may include an off-base course or training exercise, an overseas mission such as Afghanistan, or serving in the event of a national disaster either at home or abroad. For an assignment to be classified as a deployment, military personnel must be separated from their loved ones for thirty days or more. Thus, deployment to Afghanistan not only separates families in excess of thirty days but also places loved ones in high-risk war zones, leaving the family living with war. As a result, frequent deployments coupled with the current war scenario have caused a shift in stress levels and fear for loved ones. But in advance of deployment, this book looks at relocation because of its historic significance.

In the Canadian civilian realm, many families move more frequently now than at any other point in history. Often these relocations are made for what could be labeled as status reasons. Justification for making these shifts can include climbing the social or corporate ladder, upsizing or downsizing to reflect family changes, or responding to an ethnic change in one's neighbourhood. An additional reason for civilian relocation is due to the necessity for employment opportunities and career advancements. In the Canadian military realm, many of these same reasons apply to issues like changes in rank (social status) or family size—often these are accommodated by moves within a base. However, career opportunities or rank changes may also result in being posted to another base. The military's previous pattern had been to post individuals (and thus their families) to a different base every two or three years for a variety of reasons. Today, the pattern tends to be toward fewer postings. Yet of the families I met during my ministry in the CFB Petawawa area, over half have since been posted to another base in Canada. As a result, I conclude that postings continue to be a stress factor for many in our CF.

Postings are a lifestyle issue in the military; one we absolutely cannot overlook in the local church. For the past fifty years or more, this issue has been the number one cause of stress in military families. Civilians who have moved frequently will be able to identify with military families on this lifestyle issue. Many of the comments found in reading on the topic of postings pointed to a variety of difficulties.

First, there is the difficulty of leaving behind family, friends, doctor, school, church, and other areas of one's social life. My assumption here is that some family members reside adjacent to the present posting, but none may be available in the surrounding area of a new posting; close proximity to extended family is not a military concern. Added to the difficulty of leaving so much behind is the corresponding problem of making new friends and finding a doctor, dentist, church and other social outlets in order to resume family life. Today, with the shortage of doctors and other professionals, these concerns are becoming increasingly important. At the same time, they cannot be limiting factors for growth and development of either military or civilian professionals.

Second, there are always myths connected to postings, such as the doctor shortage; this may contain some truth but circumstances vary between locations. Another myth focuses on the difficulty of finding a new friend or friends—a myth Christians could nullify. We know relationships with family and friends are difficult to maintain across the great expanse of this land, especially when coupled with infrequent visits. Therefore, the Christian civilian community should encourage interaction with our troops and attempt to reach, at the very least, Christian individuals and families. Churches should be a starting point in the search for friends and fellowship. If the local church fails to understand the military lifestyle, Christian military personnel may attend and yet never develop deep, meaningful friendships. Proverbs 27:10 offers this advice: *"Do not forsake your friend …and do not go to your brother's house when disaster strikes you—better a neighbor nearby than a brother far away."* Churches that fail to understand the military forsake Christian military friends and obligate them to rely on distant family relationships. Scripturally, we are called to not forsake a friend, which is what we do by failing to understand others. Rather than let them focus upon myths about a posting, we should help dispel those myths and encourage families to consider the opportunities possible from the experience. Postings raise stress levels as every move will—military or civilian; but stress is not always negative. When forced into a stressful life crisis, how one responds is a personal decision. Admittedly, military families may face crises with more frequency than most Canadians, but perhaps we can take courage from the popular interpretation of the Chinese characters for crisis. On their own, 危机 symbolize danger and opportunity. There will be danger associated with leaving relationships behind; but in the process, opportunities abound for those willing to reach out and grasp them. In other words, personal attitude is a key to life's experiences—especially the experience of being posted. So how does one cope with the continual postings required by the military? Rather than attempt to answer that question, it is appropriate to remember there are no one-solution-fits-all answers. Nevertheless, a starting point is to encourage and promote an optimistic attitude. Couple optimism with the old saying: "Believe nothing you hear and only half of what you see" and you have a great start for coping with postings. If one listens to others, they will find themselves dreading the move, which only serves to ruin their state of mind before leaving and their attitude upon arrival.

Third, the idea of putting down roots is one most people may not think about. But the truth is military families are experts at moving, quickly making a house into a home, finding the essential community services and so forth. However, they also know in two or three years the process may well be repeated. In other words, there is a floating or liquid aspect to family life; they lack a deep-rooted community connection. As a former dairy farmer, I appreciate the three Rs of life: Relationships, Roots, and Routines. God's socializing plan for His creation is found in the herding instinct of domestic animals, packing mentality of wild beasts, flocking of birds, and schooling of fish. Dairy cattle in particular thrive as part of a herd even though they are subject to an established hierarchy that gives each animal their own social position. It is part of the social fabric in the life of a dairy herd, and new additions quickly discover their relationship status within this social structure. In addition, anyone who has milked cows understands the importance of routine—twice a day, seven days a week. Humans have also been designed by God to live in relationship with Him and others; additionally, we are genetically designed to put down roots and develop routines. Relationships were initiated by God in Genesis 2:18 with the words: *"It is not good for the man to be alone."* This reference is not directed solely at the relationship between husband and wife, because it includes societal interactions as well. In Genesis 1:28, we read of God commanding man (male and female) to put down roots through this commandment: *"Be fruitful and increase in number; fill the earth and subdue it."* To do so requires putting down roots and remaining in one location. Thus, God continues: *"Rule over the fish of the sea and the birds of the air and over every living creature that moves on the ground."* This requires establishing a routine with those creatures.

For human beings, the three Rs are God-ordained. Regrettably, they are knocked out of balance for military families with every posting. Each time families are posted to a new location, they resemble a new addition to a dairy herd—they must find their place in a new social structure. It may seem like a minor issue in our high-technological, jet-set society, but roots continue to be valued, especially when starting a family. When a family moves they remain in relationship with one another and they soon establish new routines. On the other hand, rootedness to an area is not so easily established. A rooted family is comparable to a maple tree with strong, deep roots penetrating to great depths in the soil that only a tornado can up-root. A military family lacking roots resembles a cedar with shallow roots close to the earth's surface, easily toppled by strong winds. Roots woven into the social fabric of a community are priceless when the storms of life hit; these are often lacking, missing, or unavailable to military families. As an example, one tornado-style life storm for me was the divorce that uprooted my family. In the aftermath, I struggled with moderate depression and suicidal thoughts while becoming re-rooted; but support from my biological family and community of faith served as deep, stabilizing roots. I was fortunate to have these deep maple roots. It is important for Christian civilians to realize their responsibility to protect and tend the spiritual and social roots of every member of God's family, especially military members.

There are a number of other posting issues civilians may overlook. I will focus on each briefly, continuing the numerical sequencing. Fourth, some may question the purpose of multi-postings but the concept is valid. Staying in one location too long leads to repetitive approaches and stagnant thinking, and stymies creativity. It certainly does not motivate originality. When one considers biblical characters such as Abraham, Jesus, and Paul, it soon becomes evident they were constantly posted to new locations in God's service. In fact, remaining in one location leads to more of the same old, same old or worse, as David's affair with Bathsheba demonstrates.

Fifth, some raise the issue of the work involved with frequent moves. I tend to agree with Harrison and Laliberté[84] who object to the moving burden being laid on a wife. Although, I believe this applies in the majority of moves both military and civilian. Often but not always, packing, organizing, and preparing for a move become the wife's responsibility. As unfair as it may seem to saddle one partner with the task, it also points to a flaw in the communication and planning process between a husband and wife. Thus, postings raise a problem mentioned in the first chapter—communication breakdown.

Sixth, is it possible earlier generations held a different attitude than today's military spouses? In all likelihood, they probably did. Society in recent history appears to have adopted the "What's in it for me?" attitude. This question of attitude surfaced during my interview with Collier:

> In some ways, Dave, and this is not a popular opinion to have but I have been at this a long, long time—but let me emphasize *'in some ways'* I think we are doing too much for our families and we're doing too much for our soldiers and I know people would react strongly to that, but…I said this in my second book…we are teaching them to depend on the system—not depend on themselves or each other—what's going to happen twenty-five years down the road…?[85]

Collier goes on to say she does not think our soldiers will be prepared for the exit gate when that time arrives. In my opinion, what she is describing is an attitude change—one that has shifted from reliance on self and others to system dependency, or a "What's in it for me?" attitude. This is not an attitude exclusive to our military, but one I believe permeates Canadian culture today.

Seventh, the question of cultural issues as a stress factor during a move should be considered. A Francophone wife being posted to an Anglophone part of Canada would be an example. No doubt this requires

---

[84] Deborah Harrison and Lucie Laliberté. *No Life Like It: Military Wives In Canada.* Toronto, ON: James Lorimer & Company Ltd, 1994
[85] Collier Interview. 10–11

cultural and linguistic adjustments, but we must not forget Canada is a bilingual nation and the reverse, English to French, is equally stressful. Author Jeanette Russell[86] mentioned crossing a multitude of cultural barriers, and in so doing discovered excitement and adventure with each new posting. The additional reasons listed above—logic of postings, posting demands, attitude, and culture—may seldom be considered by civilians. Civilians tend to move for status or employment reasons, not as a mandatory requirement of employment.

The final issue is one close to the heart of Collier. This paragraph focuses on the girlfriends of men in our military. During my interview with Collier, she expressed a real concern for these young women. She believes a successful military marriage is based on communication, trust, and understanding the lifestyle. In the final analysis, the ability to cope with posting after posting after posting may depend on a number of issues, but it certainly has a great deal to do with starting off with a solid foundation—one established early in the relationship. From personal interaction, the truth of Collier's observation is evident; her three keys—trust, communication, and lifestyle knowledge—are the essential ingredients for a happy and satisfying military life together. As I reflect on my marriage failure, Collier's three keys take on added significance; agriculture, like the military, has a unique lifestyle. We will look at the topic of marriage more fully in chapter seven. But for the development of a ministry to military families these keys are fundamental. It must be a ministry that is capable of trusting God fully, communicating effectively to both the military and civilian community, and promoting lifestyle knowledge. A civilian ministry to our military must be prepared to accept and cope with postings; simply put, they are part of the lifestyle military families have accepted. Thus, effective ministry will require taking a positive approach to this revolving door that can affect congregational life and military ministry programs.

Once again, as we prepare to be posted to the next lesson, a study of the challenges associated with deployment, we'll take a moment to focus on biblical parallels related to this issue of postings.

---

[86] Jeanette (Shetler) Russell. *We're Moving Where? The Life of a Military Wife*. Renfrew, ON: General Store Publishing House, 2003

Biblical Parallels

In the Old Testament, the primary example of a family constantly being posted is found in the life of Abraham and Sarah. Their journey through life has been summarized in the poem that preceded this chapter. In reading the poem, you walked with the family for the final one hundred years of Abraham's life, hopefully realizing his flexibility and willingness to obey God. His son Isaac and grandson Jacob (aka Israel) were also frequently in transition for one reason or another. Moses led the Israelites out of Egypt and to the Promised Land, which they rebelliously refused to enter. As a result, they spent forty years wandering in the wilderness, during which forty moves are recorded; these are listed in Numbers 33:1–49. Jumping forward, we find David relocated for a different reason; he was being pursued by King Saul. David's multiple postings are the result of trying to keep one step ahead of Saul, who wanted to terminate his life. Other biblical characters moved as well, but far less frequently. For example, Daniel was exiled to Babylon and Jonah was sent to Nineveh. However, their relocations resemble more of a deployment by God than a posting.

In the New Testament, there are two primary examples of posting parallels to be considered. The two individuals on whom we have the largest volume of knowledge in terms of lifestyle are Jesus and Paul. In the last three years of His life, the ministry of Jesus was constantly in motion as He moved from Capernaum in the north to Bethlehem in the south. He crisscrossed the area, which is about one hundred and fifty kilometers long by forty kilometers wide. For easier visualization, His life and ministry was confined to an area approximately the size of Prince Edward Island (PEI). But these are all approximations and certainly PEI comprises a very different landscape. Jesus was posted almost daily. In fact, His life may more closely parallel that of a homeless person on the street than a member of the military. In Matthew, we have a record of how Jesus lived:

> *Then a teacher of the law came to him and said, "Teacher, I will follow you wherever you go."*
> *Jesus replied, "Foxes have holes and birds of the air have nests, but the Son of Man has no place to lay his head."* (Matthew 8:19–20)

From these words, we can see that Jesus ministered as a transient, moving from place to place, dependent on others to meet His needs. Jesus lived out a thought Paul later included in his letter to the Philippians: *"And my God will meet all your needs according to his glorious riches in Christ Jesus"* (Philippians 4:19). The Son of God lived in total dependence on His heavenly Father. But bear in mind, many of the luxuries we take for granted today in North America were unknown in the days of Jesus. They are not only available to us, but many are necessary for human survival in Canada.

We now turn to examine Paul's life. Because he was quoted above, it should be noted he not only wrote or talked about living in total dependence on God, but he also lived it. Paul walked the talk. As an Apostle to the Gentiles, he travelled much of the known world of his day. He received postings which took him from Jerusalem through western Turkey and eastern Greece and back to Jerusalem—three missionary journeys. His final mission was aboard a ship that carried him to Rome. The third journey was by far Paul's most extensive, by my estimate approximately 4,200 kilometers over a span of nearly five years. Because his three missionary excursions built upon one another, a descriptive Canadian parallel might be three attempts to reach Winnipeg from Ottawa. Based on Ottawa being about 2,200 kilometers from Winnipeg, his first journey compares to going from Ottawa to North Bay and back; his second Ottawa via North Bay to Thunder Bay and back; and finally Ottawa via North Bay and Thunder Bay to Winnipeg and back. His postings in each location along the way lasted from a few hours or days to a twenty-seven-month ministry in Ephesus. Incidentally, many of his posting were terminated when Paul was run out of town. There is no point attempting to document Paul's every posting during his roughly thirty-year ministry. However, it is reasonable to assume he made at least twice as many stops as Moses. Of course, Paul did not have an entourage of a million or more people and their livestock accompanying him.

It is difficult to parallel postings issued to God's tactical unit with those given to military troops today. The following is a sample of some differences between these two groups. Numerically, fewer members of God's army moved, in comparison to current numbers in the CF. In other words, as Christians we should not expect to be constantly posted to another community or even congregation for that matter. However, neither should we become frozen in one location. Additionally, many in the Old Testament when posted left nothing behind because family and livestock moved together as a unit. By contrast, military postings today entail loading a truck with personal furniture and household items before heading for the next CFB. These moves result in leaving friends and occasionally family behind. In the New Testament, Jesus and Paul travelled with minimal possessions, but they also left family and friends behind, although both had a small core of followers traveling with them. Another difference is the issue of doctors, dentists, churches, etc.—they were simply of no concern for those in God's tactical unit during the pre- and early church period. Likewise, roots were not a concern because, as already noted, people simply pulled up and went with you.

The circumstances vary, but disruption of family life is a feature in both biblical and military postings. To conclude, the intent of this section has been to draw attention to existing parallels between families called to serve in God's tactical unit and those called to serve in the CF. Members of our CF maintain a degree of portability, and God's people must do likewise. Long-term ministry in one location can result in a lack of productivity and creativity along with stagnation and a familiarity in community relations, which could become detrimental.

We are now prepared to look at the number one stress factor today—deployment, the subject of our next lesson. The poem on deployment is provided to point out some issues wives are left to cope with during a deployment and portrays the macho role played by some men, a role often nothing more than a front. I created this non-scriptural piece to serve as a humorous break. It is not intended to put down women or stereotype our soldiers.

Commence Firing!
SIX

BALLAD OF A REDNECK MALE CHAUVINIST
Now to some, this poem might offensive be,
But it's just a country boy's humour, you see.
Heigh-ho! Heigh-ho! It's off to deploy we go,
If need be to sacrifice our all; we don't know!
Yes, 'let the fun begin' is the song we do sing.
This we've trained for, not some other thing.
At two a.m. by bus to get the jet we'll take
No need for mushy goodbyes for pity's sake.
A carefree bunch 'board we'll all climb
No need to budget down to my last dime.
The wife she don't understand this passion,
But an adrenalin rush does suit my fashion.
So home alone with kids and pets she'll be.
What concern do you suppose that is to me?
The Rear Party guys are there for her call
If toilet leaks and water runs down the hall.
The snow can blow and build a mile high
I'll be way off somewhere warm and dry.
Or is it the grass she'll now have to mow
'stead of sittin' on her butt, lazy so and so.
Or maybe the car, there her problem will be.
Guess she can sit at home dreamin' about me.
What if many a night she spends home alone?

Nightly now she spends to Mom on the phone.
A sick kid or to the hospital must run—
Why, that's all part of a mother's fun.
My cheque, why she'll spend it all,
No doubt shoppin' and havin' a ball.
But I'll be far off with grenades in hand
And with gun I'm ready to take my stand.
Or 'top a LAV there for miles to see
'hind a grenade launcher smilin' with glee
I'll fire at them there enemy guys—no fear
Picturin' 'em as just an out-o'-season deer.
No antler trophies there'll be to fetch home
Still o'er those battlefields I'll freely roam.
And when the fight, we have her won,
I'll come home t' the wife and son.
But if by chance she did fly the coop,
With guitar on lap I'll sit on the stoop,
Or load hound 'n gun in our old truck
As we scout 'round for some old duck.
And thus to stay or go she's free to do
Won't see this redneck sit and stew.

COMMENCE FIRING!

The order "Commence firing" given during base training does not stimulate an adrenalin rush to the extent of this order in a war zone. Deployment places troops on the front line, which is less defined in modern warfare than at any time in history. In Afghanistan, the Tim Hortons outlet on the base at Kandahar had a contingency plan in place in case of incoming mortar fire. The fact this plan had to be implemented illustrates the difficulty in defining front line location.[87] These stress factors are not new to deployments, particularly if you consider some of the similar problems that faced participants in World

---

[87] Jennifer Jones. "Tim Hortons in Kandahar, Afghanistan: An insider's view." Available from http://www.canadianliving.com/life/community/tim_hortons_in_kandahar_afghanistan_an_insiders_view.php; accessed March 20, 2013.

War II. Convoys across the Atlantic faced the constant threat of German U-Boats and aerial bombardment that created an inevitable tension among the crews. In today's war theatres, it seems this tension has only increased. Incoming missiles are launched at high speed from concealed locations with little or no warning of incoming attacks. The problem of identifying friend from foe has increased exponentially with the rise in terrorism and use of guerilla tactics in the Middle East. Thus, there is even less opportunity to detect enemy action and evade it, despite the rapid advances in technology since WWII. The word deployment has already been defined as an assignment of military personnel that separates them from loved ones in excess of thirty days. Deployed CF members live six-month stretches in a tense environment, uncertain of what each moment may bring. This causes adrenaline levels to be constantly elevated as it did in WWII when the duration, location, and even mission often changed rapidly or were completely unknown to the member. For loved ones living with such uncertainty far away, hampered communications cause much anxiety. This was clear in the past, and continues within present-day military families despite the advances in modern technology. In World War II, written correspondence could be censored, delayed by months or even lost completely. The speed of written communication has improved thanks to the Internet, but remains a challenge for military members due to infrequent access to computers and connection problems in the field. Long-distance phone calling also helps to solve many of the communication problems of the past, although it has its own problems due to time zone differences, voice lag, and static on the line. Most civilians probably grasp and empathize with the stress military families endure when a loved one is deployed. What may not be as obvious to civilians is the degree of stress military families encounter in pre-deployment and post-deployment.

As an overview, deployment stress does not begin when the trooper steps aboard a flight. Nor does it end six months later upon arrival home physically safe. Family stress begins well before the actual departure and continues after a spouse returns, sometimes long after. What civilians may not appreciate is the fact that there are three phases to deployment stress. The emotional cycle begins with pre-deployment stress (knowing a spouse is leaving), moves to higher stress levels during the actual period of deployment (concern for each other and daily coping apart), and concludes with post-deployment stress (readjustment upon return). During the actual deployment, there is an additional period of secondary stress as well. This occurs when the spouse returns home on leave, generally for two weeks. Many military families consider the period from mid-deployment leave to deployment termination as an arduous task because this "here today, gone tomorrow" stage temporarily upsets emotions stabilized during the initial months of separation. Post-deployment stress issues have been viewed as a major concern only in recent years. Canada has failed to keep pace, especially with the USA, in acknowledging the need for treatment. We lag well behind the USA, especially in diagnosis and treatment of PTSD, which varies by individual. A brief history of warfare trauma response shows it has been labeled by different names in the past. A complete, concise description of past trauma identification is offered by Adsit:

> PTSD has been called by many terms over the centuries, which makes it clear that it's a disorder not unique to modern wars, but common to all wars. During the latter half of the 1600's, the Swiss observed a consistent set of symptoms in some of their soldiers and called it "nostalgia." German doctors of the same period used the term *Heimweh* and the French called it *maladie du pays*. Both terms are roughly translated as "homesickness." The Spanish called it *estar roto*, "to be broken." During America's Civil War it was called "soldier's heart." They called it "shell shocked" in World War I, "combat fatigue" in WW II, and "war neurosis" during the Korean war. In the 1970's they coined the phrase "Vietnam Veterans Syndrome."[88]

As our troops return suffering with PTSD to varying degrees, Adsit has coined a phrase worth remembering: "Post-traumatic Stress Disorder

---

[88] Rev. Chris Adsit. *The Combat Trauma Healing Manual: Christ-centered Solutions for Combat Trauma.* Newport News, VA: Military Ministry Press, 2007. 22 (Author's emphasis)

is a *normal* reaction to an *abnormal* event."[89] There is nothing normal about what troops experience in a war zone, therefore there is no such thing as an abnormal reaction. We'll focus on PTSD later; however, its impact on individuals and their families makes it necessary to draw attention to it in this overview.

Deployment does not always involve a tour of duty in a theatre of war; past actions have included assistance to both international and Canadian communities. Internationally, troops have been deployed on peacekeeping assignments and to aid in relief efforts. At home, troops deploy for training maneuvers and to aid with natural disasters such as ice storms, flooding, hurricane damage, snow removal, etc. When a call for assistance is received, our troops respond quickly and deploy. These peacekeeping and relief efforts often involve a family separation in excess of thirty days. The CF is not just Canada's war machine!

Turning to an examination of the three-phase deployment stages, one soon learns there is considerable material available on the topic. Military Family Resource Centres (MFRCs) have programs and publications to assist families through the emotional experiences of a deployment. Their programs include children's events, deployment workshops, support networks, special monthly activities, and more. "The 7 Stages of the Emotional Cycle of Deployment"[90] is a beneficial three-page summary breaking the three-phase deployment into seven stages. Leave at the mid-deployment phase noted above is omitted or ignored; however, it does repeat the deployment phase stages. One should keep in mind this cycle of emotions applies primarily to the spouse remaining on the home front; those deployed may have less time to think about emotions. They do, however, have anxiety and concerns for family at home, which will be pointed out later. The seven stages of the three-phase deployment cycle are:

    A. Pre-Deployment: Anticipation of Loss
    B. Pre-Deployment: Detachment and Withdrawal

---

[89] Adsit. 23 (Author's emphasis)
[90] *The 7 Stages of the Emotional Cycle of Deployment.* Source: Jennifer L. Hochlan for LIFELines. A publication available from the Petawawa Military Family Resource Centre, n.d.

C. During Deployment: Emotional Disorganization
D. During Deployment: Recovery and Stabilization
E. During Deployment: Anticipation of Homecoming
[two weeks of leave repeats C., D., and E.]
F. After Deployment: Re-negotiation of the Marriage Contract
G. After Deployment: Reintegration and Stabilization

Elisabeth Kubler-Ross developed her eminent five stages of grief in her 1969 book *On Death and Dying*[91] with others expanding on her list. The stages of grief are frequently applied to other emotional upsets such as divorce, major property loss, financial disaster, etc. As colossal an emotional upset as deployment can be, these stages of grief do not correspond well in this case. There may be a couple of reasons for this: first, conditioning resulting from the knowledge that deployment is a term of employment with the CF. Second, there is hope from knowing support from an MFRC and other wives exists, even if one does not utilize it. Nonetheless, emotions during all phases of deployment obviously run very high. It is imperative Christian local civilian communities be aware of stress and tension involved at this critical time in a family's life.

The pre-deployment phase embraces two stages. Anticipated loss begins with the base commander announcing the CFB will send personnel to a particular location. Now each base family realizes one or more family members face possible deployment. Our natural instinct is to prepare for separation from a spouse or, in the case of children, a parent. Families progress to stage two with release of information identifying the individuals to be deployed. The families now know definitely if a member or members are scheduled for deployment. This process may parallel in some ways one of two scenarios: you suspect you have cancer and then receive a doctor's confirmation; or your marriage is in trouble and then receive notification of divorce action. Of course, these are never exact comparisons, but they can still assist in relating to a similar situation. When deployment news, a confirmed cancer diagnosis, or a divorce notice arrives, the event is no longer anticipated; it becomes

---

[91] Elisabeth Kubler-Ross. *On Death and Dying*. New York: Macmillan Publishing Co., 1969.

a fact of life. Our human mechanism for coping with a pending loss takes control and family members want to withdraw from one another in preparation for the pending disconnection. While we may suspect the opposite reaction (except in a divorce), it is human nature to harden our feelings in preparation for an emotional injury. This is to be expected but not necessarily embraced. In fact, couples should resist their natural tendency and spend more time with one another as a couple and a family during pre-deployment. With the family unit together, it is an opportunity for face-to-face communication, which will soon be unavailable.

What is it like to be in the second phase and have your spouse away for six-month periods? When a spouse boards a bus or plane to begin deployment everything changes again, heightening emotions. Naturally, the deployment destination plays a role in the level of stress, because deployment to flooding on Manitoba's Red River would not be as stressful as heading for Afghanistan. Regardless of destination, the family remains separated by insurmountable distance and annoying time differences. Relationships change due to prolonged periods of absence, multiple deployments, and time alone. These are the dips, dives, and twists on an emotional roller coaster common to military families, especially spouses at home. This emotional roller coaster gives rise to divorce rates higher than the Canadian average. Without trust and deep commitment to one another, divorce may be viewed as an option to escape continuing stress and isolation in the lifestyle, although this is not necessarily the best solution. The deployment period encompasses three stages interrupted by a two-week leave in long-term deployments, creating its own unique emotional roller coaster. With leave, the stages of the emotional cycle of deployment are again repeated. In effect, family members work through a second round of emotional disorganization, recovery, stabilization, and expectancy. The phase begins with emotional disorganization or an overwhelming awareness of what lies ahead. During this stage, a spouse can be highly susceptible to depression. Many women report low energy levels and lack of motivation during this initial period. Some wives with families to care for find it difficult to maintain daily routines. Things such as not getting out of bed in the morning, allowing dishes to pile-up in the sink, refusing to tackle laundry, eating away from home rather than making meals, and other expressions of disinterest in usual daily chores. Duration varies between individuals but normally progresses to recovery and stabilization. What is demonstrated in stages three and four are part of the grief cycle mentioned earlier as a spouse moves from a state of possible depression to acceptance with its accompanying stabilization. In other words, establishment of new routines give life a sense of normality. Before long, spouse and family begin to anticipate their CF member's mid-deployment leave and a countdown to this homecoming begins. Once home, the emotional cycle for mid-deployment leave resembles stages one and two of the first phase. Some may think of this period as a second honeymoon, but most will view it as just part of military life's emotional roller coaster. Once this two-week leave is completed, phase two restarts, culminating with anticipation of a final long-term homecoming. However, the dips, dives, and twists do not end with a deployment's conclusion. Battle-weary troopers can arrive home physically intact, but mental and spiritual battles may await them.

Issues mentioned above represent some of the heartaches and challenges faced by a spouse on the home front during deployment. In addition, chapter four focused on the issue of living *with* war or its threat from a home-front perspective. Now my attention turns to the front line and living *in* war. Lieutenant-Colonel Ian Hope offers insight as he shares his experience from a 2006 tour of duty in Afghanistan:

> Risk of injury or death was shared, and I, and the sub-unit commanders, came to know intimately the fears of moving across that imaginary line that separates what constitutes safe ground and what we knew to be a dangerous place. We came to understand why soldiers kiss their personal talismans before a fight, the fatalistic acceptance of death, how to deal with that visiting shadow that popped the dreaded question in my head "will I ever see my children again?"; and how to dismiss fears and carry on with our tasks. These psychological oscillations are never present in peacetime training; they were constant companions of those who lived out-of-the-wire, and mark them

permanently from those whose relative safety precluded having to deal with that dark shadow.[92]

The deployed spouse suffers their own stress when absent from the family. Deployment to a war zone is difficult and yet numerous major and minor catastrophes occur in a family regularly while one member is serving their country. Communication may be less than perfect, but families do routinely connect during deployment. As would be expected, home problems are discussed and may subsequently weigh heavily on a spouse's mind when they can do little to assist. Thus, the potential exists for military responsibilities to be compounded by home-front distractions. I am not suggesting a communication ban, because troops need and have a right to know about affairs at home. Complicating this dilemma, however, is the frequent need for split-second decisions by troopers in a war zone. Slight distractions such as problems at home can potentially jeopardize everyone's safety. Thus, concerns over family could prove to be fatal. As a result, Christian civilian communities possess indeterminate potential for saving lives by reducing stress on military families to the benefit of those defending our nation. "The straw that broke the camel's back" is an aphorism applicable to the situation because civilians can potentially remove straws by doing seemingly unimportant or insignificant tasks.

Civilians cannot begin to understand the variety of situations faced by our troops, making it prudent to err on the side of maximum stress relief at home. Occasionally, the split-second decision making required has an equally dangerous parallel at the opposite end of the scale—that of lengthy hours of boredom. In private discussions, individuals have shared their anticipation about getting home and missing their families, particularly during boring periods of a tour. However, often as boredom sets in they are returned to reality without warning by gunfire and unbridled panic! This serves to illustrate the contrasting circumstances and pressures soldiers endure. Boredom allows time for reflection about home-front complexities, leading to increased anxiety and stress. The result is boredom that could be deadly if it causes hesitation before

---

[92] Hope. 151–2

pulling the trigger in a moment of sudden danger. Thus, split-second decisions or prolonged periods of boredom are influenced by stress and anxiety for family on the home front and both are equally dangerous.

The military suggests a third and final phase of two stages amounting to a single focus: stabilizing family harmony and balance. This period encompasses arranging priorities, weighing individual responsibilities, and harmonizing give-and-take within family life. It is a time of adjustment for each member of the family, and includes both parents and children. Stability is obtained relative to individual family dynamics, as is stabilization in the phase classified as During Deployment. Often, family life will reach equilibrium or a sense of normality without immediate conscious awareness by its members.

So what is it like for the troops who arrive home safely? Does it become a second honeymoon? Possibly, in some regard, but relationships suffer change through separation. No relationship can remain unchanged after a prolonged separation, because individuals grow and change with time. Think of a teen leaving in September for college and not returning until Christmas; there will be significant change in teen and parents. It is no different during deployment. A couple can be drawn closer together or pushed apart through separation. Each situation demonstrates individually unique characteristics and is beyond prediction. Unpredictability is the only predictable factor in military life.

With the emotional cycle involved in deployment, this question may surface: how do individuals and families cope? Fortunately, human beings are capable of coping with prolonged periods of upheaval in both our individual and family life. For example, when devastating natural disasters like earthquakes, hurricanes, tornadoes, strike, people will develop coping strategies. Individuals processing divorce, bankruptcy, death of a loved one, or other emotional upheaval find inner strength to cope. In time the damage to our emotional foundation is repaired and stability is rebuilt into one's life. How we cope is best summed up by echoing a slogan for a brand of sportswear; we just do it! Earlier, I noted a comment from Ogle suggesting deployment to a theatre of war is not a sports event for participants; neither is it for families viewing things from the home front. However, deployment does require an athletic style of flexibility, strength, and tolerance before, during and after extended periods of separation. Undoubtedly, deployment belongs at the top of today's military family stress list. Furthermore, an understanding church can aid in removing the burdens of stress from our military. Deployment is not new; it has been a factor in our Christian heritage since the Fall of Adam and Eve, their departure from the Garden of Eden to face the temporal battles we continue to fight even today. In addition to our founding parents, deployment parallels are abundant as will be discovered in the biblical parallels segment.

Biblical Parallels

The earliest recorded deployment can be viewed as the exit of Adam and Eve from the Garden. This could be followed by Noah and his family sailing away from a corrupt civilization at God's command. In the early history of God's tactical unit, a third is Abraham's deployment from Ur already discussed in the chapter on postings. These examples tend to more closely parallel Puritans departing Europe for North America. For Adam, Noah, Abraham and their families, it was perhaps more about embarking on a new way of life rather than a deployment; clearly the individuals involved left with no possibility or intention of returning. Another parallel coming nearer the mark is Joseph, as his brothers sold him into slavery, placing him in Egypt. A series of events there resulted in Joseph's elevation to second in command to Pharaoh as the country prepared for a famine. This seven-year crisis devastated the known world at that time. During the famine, Joseph's brothers were forced to purchase grain from Egypt in order to survive. Unbeknownst to them, this meant dealing with Joseph. In time, he revealed his identity to his brothers with these words:

> *Then Joseph said to his brothers, "Come close to me." When they had done so, he said, "I am your brother Joseph, the one you sold into Egypt! And now, do not be distressed and do not be angry with yourselves for selling me here, because it was to save lives that God sent me ahead of you...But God sent me ahead of you*

> *to preserve for you a remnant on earth and to save your lives by a great deliverance.*
>
> *So then, it was not you who sent me here, but God."* (Genesis 45:4–5, 7–8. Joseph's sentiment is repeated in Genesis 50:19–21)

Regardless of temporal circumstances surrounding the trip to Egypt, for Joseph it was a spiritual deployment directed by God. Joseph did not return to his family and home as is the practice today with deployment, but his entire family was posted to Egypt and there reunited with him. However, this is not the end of the story, as four hundred and thirty years later God posted the nation of Israel from Egypt to the Promised Land under the leadership of Moses. Thus from Abraham came the son of promise, Isaac; from him, Jacob, whose name was changed to Israel; from him twelve sons and the tribes of Israel; from the tribes, the nation which eventually took possession of the Promised Land. This account parallels a greater deployment started but yet to be completely fulfilled. God the Father loves the sinful inhabitants of this world enough to deploy His only son to earth. Jesus Christ completed His initial assignment on earth and celebrated a homecoming. He also left Apostles (disciples) to continue His ministry. He poured out the Holy Spirit upon them and all who would believe. His disciples were founding fathers of the Church. As the Church, we are the Bride of Christ destined to inherit Heaven's Promised Land after Christ's second coming (or deployment).

Other Old Testament accounts also in some ways parallel deployment. The twelve spies sent out by Moses, mentioned in chapter one, and the two spies dispatched by Joshua referred to in chapter four represent individuals deployed for a strategic military purpose. An interesting aspect of Joshua's deployment of spies is the salvation of Rahab; as noted in chapter four, she appears in the earthly family line of our Lord and Saviour. Later, Elimelelech and his family deployed to Moab, and following three family tragedies his wife Naomi returned to Bethlehem accompanied by her daughter-in-law, Ruth. In chapter three, Ruth was noted as a member of Jesus Christ's earthly line. Once again in chapter three, Esther's history was paralleled to a boot camp experience and it is noted again because her exile from Jerusalem to Susa represents a deployment from which she did not return. She became a secret agent as queen, playing a key role in saving the people of Israel from a plot to annihilate all Israelites. Although God is not explicitly mentioned in Esther, His hand is evident throughout the book. A final Old Testament example to consider is Daniel, a man who attended boot camp. Daniel was exiled from Jerusalem and positioned to influence Nebuchadnezzar and Darius. It was his exemplary faith and faithfulness that prompted these two rulers to offer praise and glory to God (Daniel 4:34–37; 6:25–28). His deployment, as with Esther's, had more to do with spiritual battles than temporal warfare. In the Old Testament, God's action in the accounts above can be viewed as a form of deployment designed to fulfill His purposes in history. They were all carried out by members of His tactical unit.

Jesus Christ's initial deployment, return, and future final deployment were referred to earlier in this section. Jesus is the CO in the battle against evil. Two additional New Testament figures can be understood as being deployed during their ministry. Paul, who was posted on numerous occasions during his missionary journeys, embarked on a trip to Rome for a hearing before Caesar. But, in light of words from God to Ananias and Paul's comments as the third missionary journey closed, it appears his trip to Rome was a God-assigned deployment. Speaking to Ananias, God said, *"Go! This man is my chosen instrument to carry my name before the Gentiles and their kings and before the people of Israel. I will show him how much he must suffer for my name"* (Acts 9:15–16). Caesar was a Gentile king ruling from Rome. Later, in Luke's report on Paul's farewell address to the Ephesian elders, he notes: *"What grieved them most was his statement that they would never see his face again"* (Acts 20:38). Earlier in the passage, Paul is quoted as saying:

> *And now, compelled by the Spirit, I am going to Jerusalem, not knowing what will happen to me there. I only know that in every city the Holy Spirit warns me that prison and hardships are facing me. However, I consider my life worth nothing to me, if only I may finish the race and complete the task the Lord Jesus has given me— the task of testifying to the gospel of God's grace.* (Acts 20:22–24)

This was to be Paul's final journey, a deployment predetermined by God to fulfill His purpose for Paul's life and ministry. It was a deployment from which Paul did not return.

A final deployment detailed in the New Testament is John's exile to the island of Patmos. During his time there, John wrote the book of Revelation, relating the final battle between the forces of good and evil. John writes, *"Then I saw a new heaven and a new earth, for the first heaven and the first earth had passed away..."* (Revelation 21:1). John's deployment to an isolated island was designed by God to provide John the solitude required for him to spiritually connect with God and write down the final outcome of humanity and creation of the New Heaven and New Earth. This includes a Promised Land waiting for all who trust Jesus Christ as Lord and Saviour.

For Better or Worse
SEVEN

THROUGH HIS LOVE

In Egypt on one hot and sticky summer day,
All the servants were conveniently away.
The master's wife devised a devious plan!
Her desire was to lay with Joseph, the man.
As he entered, she pleaded her female plight;
But he insisted her actions were just not right.
In haste from the house, Joseph did quickly flee,
Even as she called him with a loud, longing plea.
Thus she sought revenge, Joseph soon found;
And off to prison he went in chains, bound.
This makes no sense to our modern mind
But God had a plan, read on and you'll find.[93]

In another time and a different place,
David was head of the Israeli race.
On a warm spring night in old Jerusalem town,
King David upon his rooftop did stroll around;
'Who's this I spy with my flirty eye?' thought he,
As Bathsheba bathed in the moonbeams so free
Enjoying her privacy there in her own backyard,
The king viewed another man's wife off guard.
A woman so lovely she did capture his heart,

---

[93] Paraphrase of Genesis 39:6–20

>   And his lust for her simply would not depart.
>   He ordered her brought to his palace room
>   Where sinful desire in his heart it did bloom.
>   An abuse of his authority for sure by this king
>   But who could possibly know of such a thing?
>   A pregnancy would reveal it to her warrior man,
>   So a cover-up and murder became David's plan.
>   This makes no sense to our modern mind,
>   But God had a plan, read on and you'll find.[94]
>
>   Joseph was tempted by a married woman to sin,
>   But resisted, went to jail and in the end did win.
>   David succumbed to his passion, lust and desire;
>   Albeit loved and forgiven, yet he raised God's ire.
>   Two men powerfully used by God in a unique way;
>   But their actions about character have much to say.
>   Sin is never ignored by our Heavenly Father above,
>   But He forgives, as we are viewed through His love.

FOR BETTER OR FOR WORSE

Across the globe, the definition of what constitutes an "adult" varies widely, especially if you compare the specific "age of maturity." There is, however, the tendency throughout many cultures to equate adulthood with the ability to marry and have children. In our current cultural mindset, marriage is a normal part of life for those who have finished high school and not yet retired. Here in North America, there is also a cultural understanding that marriage is supposed to be for better or for worse; but in many ways that phrase also describes life as an adult. Adulthood in general is a period of independence that must be struggled through—for better or for worse. 'For better' is often viewed as having few marriage problems, gaining wealth, possessing physical and emotional health, raising well-adjusted children, and looking forward to a worry-free retirement. The exact opposite seems to be expected when we hear the words 'for worse,' and most hope to avoid such situations despite the

---

[94] Paraphrase of 2Samuel 11:2–15

fact that none of us, either as couples or individuals, can navigate life unscathed by pain and suffering. Ironically, it is these 'for worse' aspects of our lives that God most often uses to build character and faith. This theory is confirmed by taking a quick look at Job's life. We learn:

> *This man was blameless and upright; he feared God and shunned evil. He had seven sons and three daughters, and he owned seven thousand sheep, three thousand camels, five hundred yoke of oxen, and five hundred donkeys, and had a large number of servants. He was the greatest man among all the people of the East.* (Job 1:1–3)

Satan asked permission to test Job's attitude, and God consented. In one day he lost all his oxen, donkeys, sheep, camels, his ten children and all but four servants (Job 1:13–19). In a second test, *"Satan...afflicted Job with painful sores from the soles of his feet to the top of his head"* (Job 2:7). His wife's only comment was, *"Are you still holding on to your integrity? Curse God and die!"* (Job 2:9). Three friends came to:

> *...sympathize with him and comfort him. When they saw him from a distance, they could hardly recognize him; they began to weep aloud, and they tore their robes and sprinkled dust on their heads. Then they sat on the ground with him for seven days and seven nights. No one said a word to him, because they saw how great his suffering was.* (Job 2:11–13)

Job was most definitely discovering life 'for worse.' Yet in Job 1:22, we learn, *"In all this, Job did not sin by charging God with wrongdoing."* Again in Job 2:10, he replies to his wife, *"You are talking like a foolish woman. Shall we accept good from God, and not trouble?"* While the author adds, *"In all this, Job did not sin in what he said."* After his ordeal, Job's reply to the Lord reveals his strong character and deep faith:

> *I know that you can do all things; no plan of yours can be thwarted. You asked, "Who is this that obscures my counsel without*

> *knowledge?" Surely I spoke of things I did not understand, things too wonderful for me to know. You said, "Listen now, and I will speak; I will question you, and you shall answer me." My ears had heard of you but now my eyes have seen you. Therefore I despise myself and repent in dust and ashes.* (Job 42:2–6)

'For better' may appear to be an easy, appealing route to travel but as Jesus pointed out: *"Enter through the narrow gate. For wide is the gate and broad is the road that leads to destruction, and many enter through it. But small is the gate and narrow the road that leads to life, and only a few find it"* (Matthew 7:13–14). Struggles in life cannot be avoided; rather they should be embraced, endured, and learned from. It is for this reason military life can teach much to the Christian civilian community. Seldom do people suffer as Job suffered, but few recognize the extent of hardships endured by military families. In relation to Job's experiences, even military life is not as catastrophic but it certainly does exceed civilian norms. By recognizing the difficulties military life involves, *"a good soldier of Christ Jesus"* (2 Timothy 2:3) can consider it as a more current example of some of what Job might have suffered. Job placed the endurance bar at an unbelievably high level, but as fellow members of God's tactical unit we must strive to follow his attitude towards life for better or for worse. This chapter is divided into the following subsections: Dating, Marriage, Divorce, Social Structure, Stress Factors, Culture and Personal Ministry Illustrations.

Dating

Throughout history, there have been many different approaches to relationships within the military. Some armies marched to war with their entire community in tow, keeping their families at a safe distance from the action but close enough to be defended if need be. Others required soldiers to remain unmarried, and thus not attached to any specific location. There are even stories about armies hiring local women to travel with them in order to attend to soldiers' needs for companionship. Even today, relationships involving military members are inherently complicated—especially if they occur outside of a marriage commitment.

In her second book, Collier included a chapter entirely devoted to the problems faced specifically by the girlfriends, boyfriends, and those engaged to military members in Canada. For many, the military makes it exceedingly difficult to get to know the member because they can be so suddenly required on exercises, postings, and deployments. For those who succeed in maintaining a relationship, there is still the problem of fitting into the military community. Collier writes:

> They consider themselves to be part of the lifestyle but find that often both military wives and the system don't acknowledge them and so they feel left out... They feel they are on the outside looking in and they want more than that because they feel they are part of the 'family' too.[95]

For those young soldiers on base who cohabit with their partners and may even have children in such relationships, this feeling of alienation grows even deeper. The real heart issue, from personal experience, may not be feeling left out as much as a sense of guilt knowing their relationship is not God-honouring. People do not need to be Christian to experience guilt, and the social stigma once attached to couples living common-law causes increased stress on young couples. If guilt is the issue, it has potential to either harden and alienate young couples from God or draw them to Him. From the perspective of the church, this living arrangement is not accepted, condoned, or tolerated; but from the perspective of many Canadians, living together is a reality of dating—and our military is no exception. How can we resolve a situation in which a social norm clashes so strongly with church values? Billy Graham offers an answer through his evangelistic outreach. He closes each service with the hymn, *Just As I Am*. Graham accepts people 'just as they are' but he does not leave them where they are. Churches must adopt this attitude if they wish to minister to any couples entering their doors today—especially where the military community is involved. A process begins—one which moves a person beyond 'just as I am' into a relationship with God. By accepting them

---
[95] Collier, 2004. 27

into the church family, we have an opportunity to help them grow into a right relationship with God. As Collier points out, family is hugely significant for these young people who move frequently and far away from their own hometowns. If girlfriends feel they are not part of the military family, perhaps this is even truer of the church family? Churches must honestly address questions such as: how do young couples feel about the family of God? Are there reasons why military couples would want more of what the church has to offer? Do these couples, married or not, feel as if they are part of the church family? Does the Christian local civilian community make them feel like outsiders looking in? The church will reach military families only when it becomes like the *"men of Issachar, who understood the times and knew what Israel should do"* (1 Chronicles 12:32). In the documentary *Military Wives*,[96] one interviewee suggested that the only constant in her life was her children, while others without children spoke of depression and crushing loneliness. What a difference Christ could make in such situations by using a loving church!

Marriage

When contemplating marriage as a social behaviour, we see civilian and military couples sharing a common etiquette. Military personnel maintain traditional customs; but on occasion, members engage in unique ceremonies including full military regalia. We may never participate in such a ceremony, but we can appreciate the pomp and pageantry. There may or may not be an opportunity to learn from these military practices, but can marriage formalities outside North America offer learning opportunities? Starr raises a thought-provoking comment on our approach to marriage that I interpret as a challenge to the church:

> A friend from India commented about the American attitude toward love and marriage, "You marry the women you love; we

---

[96] Peter d'Entremont, producer. *Military Wives*, a CTV Documentary inspired by the book *No Life Like It: Military Wives in Canada*. Triad Film Productions, 1999.

love the women we marry." Therein lies a *profound* difference. In India, where the majority of marriages are still arranged by the parents, the divorce rate remains among the lowest in the world, at less than one percent! They start with commitment, and feelings *follow*.[97]

The implication is we in North America start with feelings, and may or may not follow with commitment. It is unfortunate, but deep commitment often fails to ensue—leading to divorce. This is not to imply we ought to adopt a policy of arranged marriages, but it should cause us to reflect on where our approach may have faltered. In particular, the church with its divorce rate parallel to the national average should pause for introspection. Some may argue that those statistics change dramatically when the comparison is made between those who do or do not attend a church, but in any case, these rates should not be found at all among people claiming allegiance to Christ and His church. Has the church failed to properly teach the meaning of love as defined by Paul? He provided this definition:

> *Love is patient, love is kind. It does not envy, it does not boast, it is not proud. It is not rude, it is not self-seeking, it is not easily angered, it keeps no record of wrongs. Loves does not delight in evil but rejoices with the truth. It always protects, always trusts, always hopes, always perseveres. Love never fails.* (1 Corinthians 13:4–8)

Before laying full responsibility on the church, the accountability for teaching every generation about married love is twofold. Past neglect must be shouldered by parents and church leaders for this apparent failure in society. As a parent, even one who has been through a divorce, you can model and teach the meaning of true love to your children. In parallel fashion, church leadership has an obligation to promote and live a life

---

[97] Judy Starr. *The Enticement of the Forbidden: Protecting Your Marriage.* Peachtree City, GA: LifeConneXions, a ministry of Campus Crusade for Christ, 2004. 202 (Author's emphasis)

of love. Both military and civilian marriages require a firm foundation. Thus, churches planning to lay a foundation for strong relationships and families through marriage seminars will find a number of resources in the Bibliography. Starr's book is but one of many, although her work has the advantage of not being strictly focused on the military, adding broad appeal in any local congregation.

Marriage and the family are core values in our Christian faith. The teaching of Jesus and Paul confirms this—especially the seventh chapter of Paul's first letter to the Corinthians. The military lifestyle is characteristically viewed as variant to church culture; nonetheless, it remains an area requiring ministry and one that can instruct on discipleship principles. If a church offers marriage counseling, it is essential for them to understand the lifestyle. Counseling without church attendance will be less effective, and churches unwilling to comprehend military life will repel potential CF members. Therefore, the church with a desire to minister to military and civilian couples must lovingly accept them at their point of entry into the church.

Divorce

Although there may be many legitimate reasons to end a relationship—abuse, for instance—it seems that in today's society, marriage is too frequently followed by divorce. Those famous in popular culture seem to marry on a whim, with extravagant weddings, and relationships that end after only a few scant months. Among the general populace, many couples decide not to continue in a troubled marriage, not even for 'the sake of the children.' Before the 1970s, the mindset was different. I knew individuals whose parents remained together aware even of unfaithfulness in the marriage. It was overlooked for 'the sake of the children,' with divorce sought once the kids reached adolescence. This was not unusual! Later, these children questioned if their parents had made the right decision. A quarter-century past my divorce, I now know that this line of questioning is futile. What is apparent and more significant about this event in my life is how God used it for His glory. Like other events in life, divorce does not call us to determine right or wrong but to grow spiritually and relationally.

In a lengthy comment on marriage problems in military life, Collier points to both sides of the equation:

> Living the military lifestyle is not for everyone. Unfortunately, most find this out when it's too late—they've already committed themselves by way of marriage. It doesn't matter if your partner sat down with you before the wedding and explained what your life will most likely be like—dealing with long separations; loneliness; needing the ability to be completely independent and yet giving up some of that independence upon his return to the family. It really doesn't sink in until the first time he's gone. That's when reality kicks in and you find yourself alone with all the family responsibilities, and no one to depend on except yourself.
>
> At the same time, the military man is dealing with all the same emotions. His frequent long absences from his home often make it difficult for him to fit back in. While I have heard from many women who are dealing with divorces, I know from my experience that many, many men are also dealing with the sad fact that they just can't be everything their wives need. But just like any other group, some marriages don't last for 101 reasons not related to the demands of the military lifestyle.[98]

The truth is that in any lifestyle, "some marriages don't last for 101 [unrelated] reasons." From personal experience, the financial struggles of farming took a toll on my marriage, but as in military life, there were other complications. A lifestyle may test a relationship but it cannot be blamed for relationship failure. The opportunities for unfaithfulness are numerous and gossip spreads quickly, making trust, honesty and communication key to successful military marriages. Of course, once communication breaks down, trust is betrayed, and lies are revealed, divorce is almost inevitable. Christian civilian communities must recognize the extreme and extraordinary stress level military couples

---

[98] Collier, 2004. 138

endure, which elevates the potential for divorce. Leadership and congregational attitudes towards divorcees and divorce counselling are a major consideration when creating a military ministry.

Social Structure

As noted previously, all humans are designed by God to be in relationship with Him and others, put down roots, and develop routines. God initiated relationships with the words: *"It is not good for the man to be alone"* (Genesis 2:18); a reference not solely aimed at providing Adam with a wife, but also in providing for his need for community. This need is so integral to our beings that one of our worst punishments for criminals involves putting them into complete isolation from others. By reaching out to the military community, the church can help those who may feel isolated or punished by their lifestyle. In fact, it is our Christian responsibility to encourage relationships and tend the spiritual roots of military members. By nourishing such roots, we can help military members become as deep-rooted as maple trees, strong enough to weather the storms of life. By creating a church community where members feel safe to attend church even if their lives are disrupted, we can help them to develop and maintain healthy routines in their lives.

There are, of course, certain social issues unique to military couples and their families in relation to a local civilian community. There are social standards within the military itself which civilians may find surprising. For example, as one moves up in rank, their circle of acquaintances may shrink significantly. To restate that simplistically, a major is unlikely to associate socially with lower-ranking personnel like privates and corporals. This problem can apply to family members as well. Occasionally, at mixed-rank meetings of wives, rank dictates to whom one may speak. Even dating can become more complicated depending on rank, as Collier shared:

> It's hard for the kids dating, you know; you have a young guy who's going to take you home [but] because there is an area of the PMQs where the officers live, and so as soon as a young guy

would go to take you home and—'Oh! Well your father is an officer'—it is a different dynamic all together for the kids.[99]

Today, however, Collier acknowledges a significant shift to greater social integration on base than many earlier books on military lifestyle indicate. Nevertheless, even in military social life, rank commands respect!

Collier understands the lifestyle better than anyone; she is married to it, she has written about it, lived with it, worked as a civilian with it, raised her family in it and is now retired from it. In my interview, she noted two types of young men in the forces. The first group are the young men who buy her book, *My Love, My Life*, because "they think enough of their wife and the job that she is doing in this lifestyle to support them and it shows—it shows they are interested in learning how the other half feels." The second group is those who instead say, "I'll only tell her what she needs to know." Collier's frustration was clear as she explained: "Those are the guys I'd like to take out behind the barn—you sit down and you listen—you know? But people are who they are and I feel sorry for those wives that have to depend on others to get information…it's a control issue is what it is—there is lots of that…it is such a different lifestyle."[100] She also drew attention to the on-again, off-again parenting style necessary during deployment and training exercises:

> Some people have said, "Well it's no different than being a single parent." Yes it is different! Because as a single parent you know for ever and a day or until you change your status that you're fully responsible but military wives have to relinquish some of that responsibility when their husbands come back and they're only back long enough—to really feel part of the family again and they're gone.[101]

---

[99] Collier, Interview. 37
[100] Collier, Interview. 25–26
[101] Collier, Interview. 26

I agree with Collier's comment. Parenting is a difficult task either as a couple or as a single parent, but this 'here today, gone tomorrow' approach necessary in military life adds to the complexity of parenting. One final comment on the grief cycle from my interview with Collier:

> Until you know who it is [that has been killed or seriously injured]—I'm walking on eggshells and I have—there is no one that I love directly over there—so if I'm walking on eggshells, my God, what is it like for the wives and the mothers? And there's the worry while you're waiting and when you find out who it is and it's not your loved one or someone you know, then you feel relief and then that is immediately replaced with guilt for feeling relief…there is so much empathy for those that are directly dealing with it."[102]

To civilians, the military represents an unfamiliar hodge-podge of social structures, personalities, parenting styles, and emotions meshing with one another even though relationships may not be deeply rooted. After interviewing Collier and studying her writing along with books by Harrison and Laliberté, Russell, Snailham and Taylor, I realize the social life of military families is a complex maze difficult for civilians to fathom and navigate. However, as members of God's tactical unit, we have a responsibility to strive to decipher military life in the same manner as we might interpret the lifestyle of other distant cultures with whom we are called to share the word of God.

Stress Factors

We have already considered a number of stress factors in this lifestyle. In chapter one, budgeting and second income issues; chapter four dealt with living with war or its threat; posting stresses were dealt with in chapter five; deployment was mentioned as the number one stressor of the lifestyle in chapter six; and so far in this chapter, the stigma of common-law relationships, divorce, and the irritations of rank segregation. Budgeting difficulties and divorce have probably touched

---

[102] Collier, Interview. 26

the life of every reader, either personally or through extended family; these are familiar areas of stress today. However, in healthy relationships, most stress factors can be handled by couples in partnership with one another. One partner is often more skilled with solving a particularly stressful situation. For example, I handled stress regarding the car or financial problems better than my former wife. However, when our children took sick, she knew the appropriate action to take and could be trusted to look after them. In addition, our parents were only a phone call away. If left alone with my daughters, chances are I could have handled an emergency. However, my stress level during the entire process would have been extremely elevated. Elevated stress levels are common to military spouses and they often deal with them over extended periods, sometimes up to six months or more. When coupled with one's spouse working in a dangerous situation while other family members are also thousands of kilometers away, the stress level magnifies again. Stress can be piled on by car troubles, plugged-up drains, sick children, and any number of other unexpected emergencies that may pop up. The primary stress factor may be a loved one working for an extended period away from home frequently in dangerous conditions. However, nerve-shattering stress comes also when dealing with daily aggravations and uncertainty over whom to call for help. It is at these times the church needs to become extended family prepared to assist in any emergency. In pastoral ministry, it is not uncommon to deal with calls about hospital emergencies, transportation requirements, food shortages, legal assistance, daycare centres, and the list could go on. The point is these crises stress individuals and families, whether civilian or military. But particularly for military individuals and families, an unexpected often short-term difficulty can create a stressful and overwhelming situation. Their separation from family support and familiar surroundings serves to compound their dilemma. Sometimes financial assistance is required, but more often they need a friend to intervene on their behalf. These are a few of the problem areas churches must be willing and prepared to deal with, especially in military ministry. Our troops and their families live with higher stress levels than their average civilian counterpart—a concept pastors and church leaders must keep in mind and be sensitive to. Are we willing to get involved, to get our hands dirty, to love the unlovable? Money alone simply cannot meet the human need and evangelize the world. We cannot pay someone else to *"go and make disciples... baptizing them...teaching them"* (Matthew 28:19–20) as Jesus instructed His disciples to practice. A church's approach to supporting our troops will speak volumes about how it applies the discipleship principles discussed in the chapter on boot camp—discipline, commitment, obedience, sacrifice, and weapon handling. These principles are demanded of Christ's followers in the Parable of the Sheep and the Goats when Jesus outlines how He will recognize the good "sheep" at the end of the age.

> *For I was hungry and you gave me something to eat, I was thirsty and you gave me something to drink, I was a stranger and you invited me in, I needed clothes and you clothed me, I was sick and you looked after me, I was in prison and you came to visit me.* (Matthew 25:35–36)

This is the attitude Christian civilian communities must fully embrace in ministry to be a stress reliever for our military. The verses also spell out in practical detail how to apply the Great Commission and Great Commandment. Besides, James writes, *"Show me your faith without deeds, and I will show you my faith by what I do"* (James 2:18). Christians are called to act as a positive catalyst for change—not to sit in a pew as an inert member of the church and society.

Culture

Because social life issues reflect on our culture, this is an appropriate opportunity to glance at culture in general. Military culture is unique from civilian culture. In civilian life, titles do not require recognition; an employee could conceivably date the company president's daughter or the well-to-do and not so well-to-do may gather for a beer at the same bar. What is portrayed as normal military protocol does not reflect in civilian social practice. The MCF Rep-Pet speaking at a men's breakfast on the topic of evangelism stated, "Civilians must recognize when they reach out to the military they are attempting E-2 evangelism."[103] The reference was to the work of Ralph D. Winter.[104] Based on Acts 1:8, Winter describes evangelism in terms of three realms: E-1 (Jerusalem or one's own culture); E-2 (Judea and Samaria or a close culture); and E-3 (to the ends of the earth or a totally different culture). The MCF Rep-Pet shared, "As civilian citizens of Canada, you are a separated culture from the Canadian soldier by an E-2 type of relationship." When one considers his comment and those of Collier, it is obvious the military views itself as uniquely different from civilian culture. As a result of extensive research into lifestyle issues, I agree with the assessment and wonder if Christian local civilian communities can understand military culture. Wondering is no excuse for Christians to rest on their laurels and stop striving to reach this E-2 group at their doorstep. The MCF Rep-Pet offered two other noteworthy thoughts:

1) "...If we leave things the way that they are right now, soldiers will continue to trickle into your churches by ones and twos..."
2) "...What I do believe is that in order to spread the Gospel effectively within the military community it has to be accepted within the military community...In order to become an effective part of the church community it has to have an indigenous church, if you will..."

As I understood the MCF Rep's comment, if we continue as we are currently, we may only see a marginal increase in our faith communities. In other words, if Christian civilian gatherings continue doing the same-old, same-old, results will be minimal. An alternative is for the

---

[103] All quotes from the MCF Rep-Pet were taped and transcribed with permission during a presentation to the Wesley Community Church Men's Breakfast, Pembroke, ON. March 7, 2008. Transcript filed with MCF Information. All Transcribed quotes used with permission

[104] The reference here is to: Ralph D. Winter. *Perspectives On The World Christian Movement: A Reader*. William Carey Library Publishers, 1999. "E" symbolizes Evangelism and the Number one of the three levels of evangelism Christians are called to engage in either personally or through mission support.

body of Christ to become proactive by working together toward an indigenous church. That is, a church originating in and belonging to a cultural group. Missionary J. Hudson Taylor of China Inland Mission is an early example of someone using the approach. He was frowned upon for "adopting native dress"[105] in his outreach to the Chinese. It is generally the approach taken with youth ministry by encouraging youth leadership and music styles. I remind readers my position is that youth and military ministries are both mission fields—mission fields of an E-2 type. Again, my understanding of the MCF Rep's comment is that the Christian civilian community should consider support for a cross-denominational, indigenous gathering of military Christians on base. This would be an independent church ministry, not the daughter of a local congregation or a division of chaplaincy programs (CPs). It would not be designed to compete with local churches or CPs, but rather to complement them. Protestant denominations have for years recognized that no one style or format of worship suits everyone. What the MCF Rep-Pet suggests is a 'church plant' that is not denominationally focused but community-driven, with MCF the group responsible. Numerous questions immediately surface about financing, accountability, community support, etc., but these are beyond the delimitations of this book.

Personal Ministry Illustrations

Once churches acknowledge the uniqueness of military culture, understand ministry with the military constitutes an E-2 type of evangelism, and recognize elevated stress levels play a key role in their lives, it's time to take steps to alleviate as much stress as possible by ministering to people in the military lifestyle. But how is all of this accomplished? I offer the following examples from personal ministry to supplement earlier suggestions on how to show support to our troops. The first example addresses pastoral ministry opportunities in general that I have engaged in. But for a pastor to carry out these duties, the full support of the congregation is necessary. The second example

---

[105] C. Gordon Olson. *What In The World Is God Doing? The Essentials of Global Missions: An Introductory Guide.* Cedar Knolls, NJ: Global Gospel Publishers, 1989. 137

is a combination of a vision God gave me for the Pembroke church and personal pastoral experience gained through ministry. Finally, an example from a ministry I participated in early in my career while part of a congregation with a very strong youth ministry focus.

During my years of ministry, I have received countless distress calls from church members and adherents. These inquiries have come from both the civilian and military population, thus providing me with firsthand pastoral experience in both situations. Frequently, these duties transpired in the more common settings of a church office, private home, hospital, nursing home, or funeral home. But occasionally, they have involved trips to less familiar surrounds like detention centres, accident scenes, and courtrooms. In the military, living half the width of Canada away from family, longtime friends, and familiar surroundings: where does one seek help in the event of a crisis or tragic event? For example, where do people turn when: their wife or girlfriend is giving birth prematurely; there is a crisis in a relationship with a pending separation; they discover one of their parents has inoperable cancer; they have been arrested for a civil offense; or their best friend dies on a mission or in an accident? In each of these cases and many, many more—who do they call? In the military, a Christian may turn to the church if they have found an accepting one; others will seek out a chaplain. But many will struggle alone or reach out to a local help agency; few marginal or non-church attendees will call on a local pastor. I contend that for many in society today the church is called upon only when all else fails. We, the church, become their last hope when we should be their first choice. As I understand Scripture, something is very wrong with this picture! What must occur? I am not an ordained member of the clergy because I subscribe to the theory of the priesthood-of-all-believers. Individually, we are responsible to serve not only as soldiers but also as priests of Jesus Christ, whether we stand behind a pulpit, sit in a church pew, or do both. This makes ministry the task of an entire congregation—the body which must provide hands-on assistance to all families through intervention, establishment of support groups, and unconditional love. The congregation responding in this fashion will hear Jesus say, *"I tell you the truth, whatever you did for one of the least of these brothers of mine, you did for me"* (Matthew 25:40). I describe this approach to ministry as practical, hands-on, and realistic; it's the approach churches should be demonstrating nationally—to Canadian citizens including those serving in our military.

A second example comes from the ten months I spent as an interim pastor with a Pembroke congregation. Late in 2007, God drew my attention to the increasing number of singles, couples, and young families beginning to attend morning worship. These New Nesters, as they became known, seldom participated in fellowship after the worship service; with busy lives and/or children, they headed for home immediately. I recognized a need for them to fellowship with one another if not the entire congregation. Thus, I made a Sunday morning announcement inviting them to meet with me mid-week to discover if such a group would be of interest. The response was overwhelming and this college-and-careers-age group known as New Nesters began. The name was selected as a play on the familiar empty nesters term; it reflected the reality that these young people were establishing their nest. The group was comprised of singles and couples, civilian and military, post-secondary school age, and those with and without children. Generally, they met in the church building, with older congregational members providing childcare in the nursery. The members set their own agenda with pastoral guidance. Initially, fellowship was the focus because this age group needed to connect with one another. The long-term vision called for an adult Sunday school class, small groups, and outreach using a variety of recreational approaches. New Nesters continued to be a strong group blessed by God long after I left as interim pastor. I believe the longevity of this ministry resulted from its effort to reach out and touch lives.

In 1988, I was part of a newly created Congregational Christian Church in Canada congregation (CCCC). It was decided the church would focus on youth ministry. Today, I see many similarities between youth and military ministry. Therefore, I believe the suggestion by the MCF Rep-Pet is a venture into ministry well worth the risk. The youth ministry at this church began in a school gymnasium in rural Ontario rented for Sunday morning worship. We hired a youth pastor who established a monthly Sunday evening youth worship service early in his ministry. This indigenous worship service still attracts Christian and not-as-yet-Christian youth from a wide radius. Today, this body of believers has a beautiful facility and God continues to bless both the youth ministry and congregation. Undoubtedly, there were risks associated with this venture, as there will be in the proposal made by the MCF Rep-Pet. But certainly, as with this outreach to youth, the risk taken for God can reap a great harvest.

I've offered these ministry examples to demonstrate the features of risk-taking and creativity. These need to become part of a Christian outreach for those close to a CFB if a military ministry is to be established. However, be forewarned! Do not tackle military ministry alone; it requires creative risk-taking in partnership with MCF members or other military Christians attending local gatherings, and preferably includes chaplaincy input. Efforts to support our military by individual congregations are commendable and needed; however, in addition, a joint trans-denominational, indigenous, military-focused Christian gathering would be an excellent risk for congregations in close proximity to a CFB.

Biblical Parallels

There is a great deal of teaching in both the Old and New Testaments focused on marriage and divorce in particular. In addition, Christian bookstores carry a wide array of material covering these topics from teen dating and sex to divorce and re-marriage. It would be repetitive to elaborate on these issues in this section. Finding biblical personalities who dealt with living life for better or for worse presents its own challenge. Although not exact parallels, we may find reflections in the lives of Ruth, David, Bathsheba, and Esther as Old Testament examples, with Joseph, Mary, Ananias, and Paul's shipwreck experience from the New Testament.

Ruth's life should be familiar by now. The reason for considering her story again here is to draw attention to the move she made from her home into a foreign culture. Ruth parallels many women who become military brides and in doing so enter a foreign culture. Ruth accompanied Naomi on a new future into a culture with unfamiliar customs such as gleaning, name and property right governances, and possibly the custom that led to her arranged marriage with Boaz. In like manner, many women marry into military life and transition from civilian culture to a new future in a military culture with unfamiliar customs. This increases stress in the already hectic situation of couples launching an independent identity and establishing a marriage relationship, prioritizing responsibilities, budgeting, etc.

David endured massive social and cultural changes as well during his early life, as detailed in 1 Samuel 16:1–23. He moved quickly from a sheepherder to being the next king of Israel. Along the way, he found himself serving in King Saul's court, battling Goliath the giant, making friends with Saul's son Jonathan, and even having to flee from Saul to avoid assassination. It should not surprise us that the shepherd anointed as future king of Israel would have experienced quite a bit of culture shock. David learned to be a shepherd, royal prince, man on the run, resident of Philistine and Moab, cave dweller, desert rogue, warrior, hero, and eventually king. In all of these cultures, David acknowledged God's supremacy, most notably through his understanding of Saul as God's anointed King of Israel. Scripture reveals David's heart:

> *He said to his men, "The Lord forbid that I should do such a thing to my master, the Lord's anointed, or lift my hand against him; for he is the anointed of the Lord."* (1 Samuel 24:6)
>
> *But David said to Abishai, "Don't destroy him [Saul]! Who can lay a hand on the Lord's anointed and be guiltless? As surely as the Lord lives," he said, "the Lord himself will strike him; either his time will come and he will die, or he will go into battle and perish. But the Lord forbid that I should lay a hand on the Lord's anointed. Now get the spear and water jug that are near his head, and let's go."* (1 Samuel 26:9–11)

Few individuals will experience the numerous stress factors that influenced the life of David. In addition to the above, he was married numerous times and technically divorced once. David's first wife was

Michal, Saul's daughter. Saul asked *"no other price for the bride than a hundred Philistine foreskins, to take revenge on his enemies"* (1 Samuel 18:25). It was his plan to see David fall at the hands of the Philistines; his plan of course failed. It is revealed later that Saul gave *"his daughter Michal, David's wife, to Paltiel son of Laish, who was from Gallim"* (1 Samuel 25:44). The relationship was restored following the death of Saul with the details described in 2 Samuel 3:12–16, but in the following account, we can see that problems of love continued throughout David's life.

Bathsheba's life tends to produce two interpretation theories from biblical scholars. There are theologians who view her as a powerful lady in David's court who enticed him into an adulterous affair. Conversely, others interpret the affair as solely David's responsibility. I share the latter point of view and interpret the incident as an abuse of David's authority. When one accepts Scripture as written and ignores historic speculation on the influence women may have held, there is strong evidence David used his power and authority to satisfy the lust of his heart. The affair is detailed in 2 Samuel 11:1–12:24. As Bathsheba bathed in the privacy of her backyard, David saw her from the rooftop of the palace and ordered her brought to his chambers. She was sent home after David had satisfied his desire, but she conceived and sent word to David, saying, *"I am pregnant"* (2 Samuel 11:5). Her husband, Uriah, was one of David's fighting men who had been deployed to fight for his king and nation. Imagine the fear that must have gripped Bathsheba when she was summoned to the king's palace late at night, her feeling of disgrace after being raped by the king, her anxiety over how this could be explained to Uriah, and the courage required to inform David she was carrying his child. David reacted by recalling Uriah to Jerusalem to report on the fighting and twice unsuccessfully attempted to have him spend a night with Bathsheba. She apparently was unaware of the mid-deployment leave Uriah had been ordered to carry out. David's failure to entice Uriah to go home and sleep with his wife prompted him to instruct the CO of his troops to set Uriah up to be killed in battle. Through this period, there is no indication David communicated in any way with Bathsheba regarding her pregnancy. Now imagine the grief that must have gripped

Bathsheba when she heard of the death of her husband, her sorrow over an unwanted pregnancy thanks to the king, the anxiety of not knowing his intentions, the social disgrace of her pregnancy, the recurring fear when she was again summoned to the palace, and the courage it took to face David. However, the second encounter resulted in Bathsheba becoming David's wife. It must be noted, in Old Testament time women were considered chattels, or property with no rights; but in any case, God knew what sin had taken place and He called Nathan the prophet to confront David. A son was born and died, David lay with Bathsheba again and she gave birth to Solomon. Prior to David's death, Solomon succeeded him as King of Israel (see the account in First Kings 1:1–2:12). Bathsheba certainly experienced the most difficult deployment any wife could imagine.

Once again, we draw our attention to a frequent visitor in this section, Esther. At the beginning of her story, it seems plausible to assume that her ordered marriage to the king would have brought many changes both for better and for worse. A full account of her action in God's service is not given, although we do know that Esther risked her life when approaching her husband, the king. Like David, Bathsheba, and those today who commit to a relationship with a military member, she had to adjust to being in a very different culture. Esther served a front-line position as she battled spiritual and temporal forces aimed at the total annihilation of the Israeli people. We can scarcely imagine the courage it must have taken for her to overcome the fear of death knowing her only weapon was the words she would speak. This was a unique and stress filled assignment, which can be studied in depth in the book of Esther chapters four and five. In such a study, we find that Esther excelled in tact and resilience, and had enough courage to face the worst challenges of her life head-on. She fought for her own life and for those of her people, and followed God's calling in her life for better or worse.

In the New Testament, we begin with Joseph and Mary, who experienced a term of duty in God's tactical unit probably more difficult than even David and Bathsheba endured. Mary is foretold of the birth of Jesus (Luke 1:26–38); Joseph learns of her pregnancy and is told by God that, *"what is conceived in her is from the Holy Spirit"* (Matthew 1:20). After Jesus was born, shepherds visited the baby (Luke 2:8–20) and Magi visited *"the child"* (Matthew 2:1–12). There is an unusual temple encounter with Simeon and the prophetess Anna (Luke 2:25–38), and it is followed by their flight to Egypt and later return (Matthew 2:13–23). When Jesus is twelve years old, he worries his parents when he stays at the Temple in Jerusalem (Luke 2:41–52), and as a widow, Mary suffers the agony of watching the Crucifixion of her first born (John 19:25–27). The trip to Egypt was similar to a posting because the entire family went and experienced the cultural differences. We can be assured there was considerable stress over the pre-marriage pregnancy, great joy from all of the visitors who came from near and far, as well as great fear for their lives as they fled to Egypt to escape Herod's attempt to murder their son. There would also have been considerable anxiety during the years in Egypt, and heart-breaking sorrow as Mary viewed the Crucifixion. This couple rode the most emotional rollercoaster of all time in what was definitely a spiritual battle. A battle that will only be surpassed by Christ's final battle revealed in Revelation.

Ananias is one of those biblical characters of whom we know next to nothing. His entire recorded ministry was directed at Saul, later known as the Apostle Paul. Saul, prior to conversion, breathed *"out murderous threats against the Lord's disciples"* (Acts 9:1). He travelled to Damascus in search of *"any there who belonged to the Way, whether men or women, [so] he might take them as prisoners to Jerusalem"* (Acts 9:2). On the way, Saul had an encounter with Jesus Christ that changed his life. Shortly afterward, the Lord called Ananias, saying, *"place [your] hands on [Saul] to restore his sight"* (Acts 9:12). In response, Ananias recounted to the Lord all he knew about this man who had harmed many of the saints. Without realizing it, Ananias was caught in the midst of a spiritual battle where God was calling him to obey for better or for worse, as Esther did. Ananias feared Saul and was filled with anxiety and stress over the prospect of visiting this murderous individual, as it was definitely not an easy or safe situation for ministry! He also provides a typical emotional reaction to God's call to serve: he immediately began to make excuses. Eventually, he does face up to the challenge of ministering to Paul;

an account of his ministry is found in Acts 9:10–19. Paul shares his encounter with this member of God's tactical unit in Acts 22:12–16, and it is sufficient to say that by facing what may have been one of the "worst" or most dangerous ministry moments for Ananias, his obedience to God's calling changed Paul dramatically and started the journey to Paul's own ministry. From such stories, churches can take courage when facing the challenge of starting what might feel like a 'risky' ministry to the military community—for it is when we obey God's call to follow him for better or worse that He will most dramatically change the lives around us.

To close, we examine Paul's shipwreck experience detailed in Acts 27. Paul's life and ministry repeatedly have been and will continue to be an example for all believers. During the shipwreck, Paul finds himself in a war against nature but also with spiritual implications. Paul resolutely set out for Rome as he approached the end of his ministry, in the same way as *"Jesus resolutely set out for Jerusalem"* (Luke 9:51) as he approached the end of His ministry. In both cases, the battle lines were drawn, setting righteousness against evil. Jesus battled with His human side—*"yet not my will, but yours be done"* (Luke 22:42)—when He prayed in the Garden of Gethsemane. Paul battled with the forces of nature as he sailed to Rome. The journey commenced on an ominous note as Luke writes, *"The winds were against us"* (Acts 27:4) and their troubles were just beginning. As Luke notes:

> *We made slow headway for many days…We moved along the coast with difficulty…Much time had been lost, and sailing had already become dangerous…So Paul warned them, "Men, I can see that our voyage is going to be disastrous and bring great loss to ship and cargo, and to our own lives also." But the centurion, instead of listening to what Paul said, followed the advice of the pilot and of the owner of the ship…When a gentle south wind began to blow…they weighed anchor and sailed… Before very long, a wind of hurricane force…swept down from the island …we gave way to it…we were hardly able to make the lifeboat secure… the men…passed ropes under the ship itself to hold it together…*

they lowered the sea anchor and let the ship be driven along. We took such a violent battering from the storm that the next day they began to throw the cargo overboard...they threw the ships tackle overboard...neither sun nor stars appeared for many days and the storm continued raging, we finally gave up all hope of being saved. (Acts 27:7–20)

In this storm, Paul experienced an angelic visitation announcing God's intentions to him, saying, *"God has graciously given you the lives of all who sail with you"* (Acts 27:24). Luke continues:

On the fourteenth night...the sailors sensed they were approaching land... Fearing that we would be dashed against the rocks, they dropped four anchors...and prayed for daylight...Just before dawn Paul urged them all to eat. "For the last fourteen days," he said, "you have been in constant suspense and have gone without food...Now I urge you take some food. You need it to survive...he took some bread and gave thanks to God...and began to eat. They were all encouraged and ate some food...When daylight came, they did not recognize the land...everyone reached land safely. (Acts 27:27–44)

This was a battle against nature fought by Paul, Luke, their companions, and the ship's crew. It was difficult, dangerous, violent, potentially disastrous with hurricane forces and without sun or stars for guidance. The men were afraid of the storm, fearful of being dashed against rocks, in constant suspense and lacking nutrition, but all landed safely on an unknown island. The specific crises may differ, but unknowingly Luke has provided readers with a vivid description of military life on the front line and on the home front.

The above examples provide an inside glimpse at life for some past members of God's tactical unit. They experienced an array of social challenges, cultural adjustments and stress in His service. In His tactical unit today, we face similar conditions, but the names, places, and crises have changed. What has not changed is that God will persevere with us to the end!

Guarding Our Own
EIGHT

THE YOUNG & THE OLD
Excellent advice there is for us to unfold,
In Scripture for both the young and the old.
Once there was a man of great strength by gee!
Moses, his name, possessed enormous energy.
He was found a babe on the Nile floating free,
And soon Pharaoh's court his home would be.
Raised for a unique task, God moved him along
To receive the Ten Commandments as his song.
So honour father and mother, a promise this holds
Lasts long as you live no matter how life unfolds.[106]
Then came a man, Solomon the smartest son of all;
Many pithy sayings he left for which upon to call.
And now listen children as your parents do teach;
Pay attention, so understanding you can reach.[107]
But grandchildren are a senior's crown of glory,
And parents, children should be proud of your story.[108]
Young men in their strength, they do find fame,
And for elderly, gray hair splendor is their game.[109]
As the Babe of Salvation to a manger Jesus came,

---

[106] Paraphrase of Deuteronomy 5:16
[107] Paraphrase of Proverbs 4:1
[108] Paraphrase of Proverbs 17:6
[109] Paraphrase of Proverbs 20:29

With God giving him the name above every name.
The kingdom of heaven to a child does belong;
The greatest there humble as a child tags along.[110]
Welcoming a child in Jesus' name does welcome Him;
Welcoming Jesus does welcome the One who sent Him.[111]
Saul hunted the Way, but Christ knocked him from his stead.
His character and name changed, Paul's letters are left to read.
As for older men in your life, do not rebuke them severely,
Rather treat them as fathers by offering advice sincerely.[112]
Older men are to be taught worth of respect and temperance,
Self-controlled and soundness in faith, hope and endurance[113]
But Peter the rock after Pentecost, he stood to proclaim
The words of Joel shared through the Spirit—his fame.
In the last days, His Spirit on all people God will pour out;
With prophesies your sons and daughters will loudly shout!
Visions for your young men will come for them to see;
While dreams to be dreamt by your old men there will be.[114]

GUARDING OUR OWN

The aphorism, "once an adult, twice a child," often comes up in discussing similarities between children and the elderly. Young and old alike live life with little or no hair, no teeth, eating puréed food, wearing diapers, and depending on others for their care. The similarities become even clearer as seniors watch grandchildren arrive and prepare for their own departure. The title of this chapter may lead one to think our military's only interest is their own 'military brats' (as kids are often referred to) and veterans. However, veterans and those who did not return provide proof that the military policy of 'guarding our own' includes Canada as a nation. This chapter focuses on military offspring and former warriors, but the ideas behind it can be applied more broadly to Canada's overall population of children and elderly. Today's child represents a sensitive, precious provision

---

[110] Paraphrase of Matthew 18:3, 19:14; Mark 10:14–15; Luke 18:16–17
[111] Paraphrase of Mark 9:37
[112] Paraphrase of 1Timothy 5:1
[113] Paraphrase of Titus 2:2
[114] Paraphrase of Joel 2:28 and Acts 2:17

for the nation's future. They should be viewed as a resource to guard, protect, and train for the long-term outlook of Canada and our children's children. The elderly of our nation have contributed in accordance with their God-given giftedness to achieve the standard of life enjoyed today by all in Canada. They have an obligation to share their wisdom and they deserve to be shown dignity and respect. By design, this book requires glancing at these natural resources and sources of wisdom in a military context. As in some previous chapters, this one will also be divided—this time into three sections: Children, Veterans, and Remembrance Day.

Children

In Chapter seven, I referenced the dating problems that may be encountered by military youth. The teen years represent a challenging period in life for them as they progress from childhood through adolescence into adulthood. At the same time, these young adults have well-developed levels of understanding and can usually relate to problems of distance, time, finances, and career responsibilities. The situation is very different for younger children, so they will receive special recognition here. An attempt to include children can be found in the work of Collier, Snailham, Taylor, Harrison and Laliberté. However, those books offer primarily an adult point of view with individuals sharing struggles regarding raising children amidst posting and deployment issues. Reading them, it is exceedingly clear that the attitude of the parent plays a huge role in how children will cope. Opinions vary widely about how well children can respond to aspects of military life. At one end of the spectrum, parents lament over their children missing out on important family connections such as growing up with their grandparents involved. At the other end of the scale, others argue very differently:

> I just don't buy it when people say they don't want to move for the sake of their kids. Our kids haven't wanted to leave any place that we've lived and now they don't want to leave Victoria and go to Ottawa.
>
> I think a lot of it depends on the attitude of the parents…I never asked my kids. They were told they were going…They

know it's a way of life. The same as in our family, going to church is a way of life. Maybe with some people, hockey is part of their life. They know they are to do it.

Kids adjust.[115]

Between the extremes, the focus generally turns to the children's pride as their parent serves and advances in the military. Later on, however, there seems to be more frequent troubles with adolescents. For those who lamented moving with their children away from families, they may blame the shallow roots as an issue while the other end may express the practical attitude that some things are simply part of life and that their children are simply learning to live with this way of life. In any case, it is a commonly agreed upon theme that teens have difficulty adjusting through the dawn of adulthood.

Parent's comments are significant and valuable to consider, but it is also important to listen to our children. In her first book, Collier used a teen questionnaire to add youth input to her book, but this does not address the years before adolescence. More recently, however, Ellis[116] spoke directly with children in Canada and the United States. She has allowed children to relate their feelings and experiences in their own words. Each account is understandably as unique as the child providing it. Although there are no observable patterns, the book does provide insight into the thinking of some military children. It merits reading by parents and is recommended as a resource for teachers of military children, employees at daycare centres serving military families and church members with military congregants. Her book is unique and represents a large cross-section of opinion, because the forty children interviewed ranged in age from six to seventeen. In total, twenty-six children and fourteen teenagers. I view the fact that sixty-five percent of the interviewees were under the age of thirteen as adding significantly to her study. Almost half of the youth were Canadian with eighteen of

---

[115] Dianne J. Taylor. *There's No Wife Like It*. Victoria, BC: Braemar Books Ltd, 1985. 51–52

[116] Deborah Ellis. *Off to War: Voices of Soldiers' Children*. Toronto, ON: Groundwood Books/House of Anansi Press, 2008.

the forty interviews; these included eleven children under the age of thirteen. These Canadian families were spread across North America in CFB Petawawa, CFB Shilo, CFB Trenton, and even one family posted to Fort Bragg, North Carolina. The material provides an excellent resource to help adults understand military life as seen through the eyes of children. No one can fully relate a child's experiences, but thankfully the kids are capable of expressing themselves if adults will simply take the time to listen. Early in life, they may lack fully developed relationship skills, but they are keenly aware of family dynamics. As an example, children notice any type of parental disunity and from that they quickly learn to play one parent against another. Their capacity to verbalize the emotions and stressors related to moves, deployment separation, divorce, or grief may be limited, but they will often demonstrate and act out their feelings. Reaching out to military children is one more approach the church can employ to demonstrate support for our troops.

Veterans

We now turn to our veterans, who have earned the right to live in dignity and with respect. I maintain that support for our veterans demonstrates support for our troops who will be the veterans of tomorrow. By writing stories, Airth attempts to bridge the generation gap between children and seniors; thus preserving the fact that: "Remembrance is important—and not just once a year. Lesley Airth's carefully crafted stories lead the youngest readers through Canada's wars of freedom by showing how children faced changes and reacted to their parents' absences."[117] Veterans are not only important, but teaching children to know and remember what they sacrificed is equally important; not only one day a year, but all year long. We must never forget the sacrifices made and still being made on our behalf. Children in a military context will find Airth's book especially interesting and full of opportunities to discover how other children adapted to a

---

[117] Lesley Anne Airth. *What We Remember*. Renfrew, ON: General Store Publishing House, 2004. b.c. An endorsement by J.L. Granatstein, former Director, Canadian War Museum

prolonged parental absence often for years, not months. The book is designed for even the youngest readers and is available in both French and English. The accounts are short, contain age-appropriate diagrams, and offer discussion questions. They cover a number of World War II events including spies, telegram notifications, and the Poppy, as well as experiences of travelling on escorted passenger ships and a child of six reuniting with his father who he did not remember. World War I is remembered through the discovery of a medal from the Great War. This book serves to connect children with veterans helping them and their parents to appreciate all that veterans have done for the nation. To remind us once again, Canada's participation in the Great War of 1914–18 changed our history forever. A Remembrance Day article stated, "In many ways, the identity of the young country was forged on those bloody battlefields"[118] of Vimy Ridge, Hill 70, Passchendaele, and Ypres. In other words, the Great War was a significant turning point in the history of our nation. We should never forget that almost 620,000 Canadians served in that war and of those over ten percent did not survive. As noted in chapter two, the Christian church in Canada on the whole did not take a pacifist stand but played a significant role in the war. Airth states in her introduction, "I believe that children—even young children—should understand how fortunate we are to live in a peaceful society, and how grateful we ought to be to those who fought for our freedom."[119] I would endorse her statement but would also like to add: may we never forget those continuing to fight for freedom as conflict persists throughout the world today.

As Christians, one of the ways we can honour those who were and are serving for our freedom is to aim to understand and be trained for military ministry. The main purpose for this book was raised by a veteran speaking at an "Introduction to Military Ministry" seminar.[120] From personal experience, this individual presented the view that our military has a very high profile during periods of war, but is seen as a drag on

---

[118] CBC article online, accessed November 10, 2009. http://www.cbc.ca/canada/story/2008/11/07/f-remembrance-day.html
[119] Airth. v
[120] "Introduction To Military Ministry"—two evening seminar held May 30/31, 2008 at Wesley Community Church in Pembroke, ON.

our national economy in times of peace. It was further pointed out that during peaceful periods military personnel tend to be treated as second-class citizens. I took special note of one emphatic statement concerning ministry to the military: "The military does not need another social club; what it needs is to be welcomed into the Christian faith and the social life of the church. The real need in the church is for training in how to deal with military families." The comment drew my attention because it resonated in my mind with an earlier claim I had read by Major General Robert F. Dees:

> ...today's military people and their families greatly need affirmation, love, encouragement, opportunities for service, and the warm accepting community that the church uniquely offers. Often this happens automatically, but many times the church does not extend compassion because they are simply not sensitized to these unique needs of military personnel, or they are somewhat intimidated by them, particularly by returning warriors who have seen so much and sacrificed so much for our nation.
>
> At Military Ministry, our prayer is that churches across our land will show a heart of compassion to our nation's military... Our prayer is that you will form an intentional, relevant ministry to the military.[121]

Hunter has written an excellent book on the Celtic approach to evangelism; thus, I want to draw attention to the parallelism between the above claims and his insightful comment:

> There is no shortcut to understanding the people. When you understand the people, you will often know what to say and do, and how. When the people know that the Christians understand

---

[121] Military Ministry Publication. *Church Guide for Ministering to the Military: An Introduction to the Bridges of Healing Ministry Including How to Provide Spiritual Care for Combat Trauma.* Newport News, VA: Military Ministry Press, n.d. 3 An opening letter from Robert F. Dees, Major General, U.S. Army, Retired, Executive Director, Campus Crusade Military Ministry

them, they infer that maybe the High God understands them too.[122]

Hunter's insight reflects these comments: "The real need…is for training in how to deal with military families" and "[the church is] simply not sensitized to these unique needs of military personnel." What this book promotes is an understanding of military families so we can train in how to deal with them, be sensitized to their unique needs, and reach out to them in the name of Jesus. It is encouraging when an evangelist like Hunter, a retired American major general and a retired Canadian soldier agree with the premise of this book. Our failure to comprehend the military lifestyle as the church or as individuals will greatly impede any attempt to reach out in support of our veterans.

The following comments serve to bridge discussion to Remembrance Day as a focus. While I was ministering in Renfrew County and researching this book, church leaders frequently quizzed me, asking, "How can we support our troops?" Dialogue soon revealed that the intent of the question was to ask, "What can we do to be more visible and effective in supporting our troops?" Today, Canadians show support for our troops by standing along the 170-kilometre stretch of Highway 401 from Trenton to Toronto dubbed the Highway of Heroes. They do this to honour our fallen repatriated soldiers. Canadians also demonstrate support through prayer, by dressing in red on Fridays to show they stand with our troops, by wearing poppies to commemorate our veterans, and by attending Remembrance Day ceremonies on November 11. Four of these provide visible evidence to show we care; but are we doing enough for our troops, chaplains, veterans, and their families?

Remembrance Day

Based on news reports, attendance at Remembrance Day services has increased noticeably in recent years. There are a number of theories as to why this is happening, but as the day approached in 2008, this idea took shape in my mind: support for veterans demonstrates support for

---

[122] George G. Hunter, III. *The Celtic Way Of Evangelism: How Christianity Can Reach The West…Again.* Nashville, TN: Abingdon Press, 2000. 20

current troops, since a trooper today will eventually become tomorrow's veteran. Veteran support opens a door to Christians from every province of Canada to demonstrate troop support. How we view our heroes of the past and how we honour and remember them speaks volumes about who we are as a nation. The following email taken from *Operation Homecoming* makes my point:

> It is a humbling experience to move about the Walter Reed complex. The gritty determination of these wounded and the support they offer to each other puts a lot of the other details of daily life in clearer perspective. Regardless of your politics or how you may feel about this war, these wounded, and the dead, are an inescapable reality. I pray to God that we as a nation don't forget the sacrifices that are being made on our behalf. From now on, Veteran's Day will be a great deal more meaningful to me...[123]

Although this may be an American expressing his opinion, similar sentiments should apply in Canada since Veteran's Day in the USA is parallel to our Remembrance Day. I took an opportunity to preach on the need to recognize and support our veterans in November 2008. The setting was a small town congregation in the central part of Ontario about an hour from either CFB Borden or CFB Meaford. The congregation was receptive, and although only one couple stopped to question me further, they saw value in the idea. I like to think others took the 'support for our veterans' idea home with them to give it some thought. New attitudes begin this way—one person or family or book at a time. A community reaching out to assist veterans will soon be known for its love.

Because of the need to reach out to veterans as well as current military members, this book ends up applying to most communities across Canada. They may not realize this, but there is a military connection in innumerable Canadian towns thanks to the presence

---

[123] Andrew Carroll, ed. *Operation Homecoming: Iraq, Afghanistan, and the Home Front, in the Words of U.S. Troops and Their Families.* Chicago: The University of Chicago Press, 2008. 357

of active and retired military personnel, as well as their families. This is perhaps most evident from the local Legion halls. Thus, respect shown to veterans today reflects our loyalty to stand with our troops and points to what they can expect in the future. Another speaker from the aforementioned seminar put it this way: "I view myself as ex-military but I cannot become a civilian again—once military, always military."[124] A similar sentiment was expressed in an email to Collier when a veteran's wife pointed out:

> Ex-military members still feel and will always feel they are military members, whether they wear the uniform or not. It's an ingrained feeling that stays with them even after they've left the CF (they never really leave, do they?). I believe the same holds true for their spouses.[125]

These may not represent everyone's experience, but many former military members are drawn to local Legion halls because of a common military bond. Therefore, this book has broad application to Christian civilian communities across our nation. If this raises your desire to offer more assistance, the book has achieved much of its purpose.

Biblical Parallels

Before turning our attention to children and seniors in Scripture with parallels to military brats and veterans, I will start by re-stating the fifth Commandment: *"Honour your father and your mother, as the Lord your God has commanded you, so that you may live long and that it may go well with you in the land the Lord your God is giving you"* (Deuteronomy 5:16). The tendency is to apply this commandment only to children, however, it is also inclusive of adult children who should still be respecting and honouring their parents. We do not have extensive detail on the lives of many children in Scripture, but we do have some knowledge of three: Isaac, Samuel, and Jesus. Respect for elders represents the other half of this chapter and a number of parallels are available. The focus will be

---

[124] "Introduction To Military Ministry" seminar, Pembroke, ON. 2008
[125] Collier, 2004. 193

on respect or disrespect shown to elders including Jacob, Moses, David, Paul, and Luke.

We should remember that Isaac was the son of promise born to Abraham at the age of one hundred when his mother Sarah was ninety years old. He was circumcised on the eighth day according to God's command (Genesis 21:4). Beyond that ceremony, we know nothing of his childhood until Abraham was instructed to offer Isaac as a sacrifice (Genesis 22:1–19). We do know *"Abraham stayed in the land of the Philistines for a long time. Some time later God tested Abraham"* (Genesis 21:34–22:1). This has caused some commentators to assume that Isaac was in his adolescent years when the test occurred. Following the test, we read, *"Some time later Abraham was told [of his brother's sons]…Sarah lived to be a hundred and twenty-seven years old. She died at Kiriath Arba…"* (Genesis 22:20–23:2). Thus, Isaac was thirty-seven at this point. Assuming the two "some time later" phrases to be rough equivalents, it would be reasonable to assume Isaac would have been in his late teens when he was offered as a sacrifice. We also know Isaac was strong enough to shoulder the wood for what was to be the sacrifice of his life. My point is to demonstrate that Abraham was no match for a robust young man if Isaac had decided not to submit to his will. The text informs us that *"Abraham built an altar there and arranged the wood on it. He bound his son Isaac and laid him on the altar, on top of the wood"* (Genesis 22:9). Apparently, Abraham and Sarah had raised a very obedient and trusting son.

Samuel was a son of prayer and faith given to Hannah after years of being unable to conceive. Prior to his birth, Hannah prayed to the Lord;

> *O Lord Almighty, if you will only look upon your servant's misery and remember me, and not forget your servant but give her a son, then I will give him to the Lord for all the days of his life, and no razor will ever be used on his head.* (1 Samuel 1:11)

After his birth, Hannah said to her husband, *"After the boy is weaned, I will take him and present him before the Lord, and he will live there always"* (1 Samuel 1:22). We read of her fulfilling the vow: *"After he was weaned, she took the boy with her, young as he was… and brought him to the house of the Lord at Shiloh"* (1 Samuel 1:24). The boy, Samuel, was raised by Eli and *"the boy ministered before the Lord under Eli the priest"* (1 Samuel 2:11). It is interesting that the only information we have about Samuel's upbringing is the contrast between him and Eli's wicked sons. Samuel was the last of the Judges, a man powerfully used by God, and the complete opposite of Eli's own sons. Samuel might be thought of as Eli's adopted son, one of whom he could be proud. Eli's natural sons died prematurely as God revealed to Samuel:

> *At that time I will carry out against Eli everything I spoke against his family—from beginning to end. For I told him that I would judge his family forever because of the sin he knew about; his sons made themselves contemptible, and he failed to restrain them. Therefore, I swore to the house of Eli, "The guilt of Eli's house will never be atoned for by sacrifice or offering."* (1 Samuel 3:12–14)
>
> Samuel told [Eli] everything, hiding nothing from him. (1 Samuel 3:18)

Eli had been warned by a man of God that his two sons would both die on the same day as a sign to him (1 Samuel 2:34); they did. Samuel honored Eli while his own sons ignored and dishonoured him to their demise.

Jesus is the Promised Son. He was born, lived, died, rose again, ascended and poured out the Holy Spirit that we might have eternal life through faith in Him. We do have some limited information about His childhood. We are told Jesus received a visit from shepherds on the night of His birth (Luke 2:16–18). Like Isaac, He was circumcised on the eighth day according to the law (Luke 2:21). On the fortieth day, Jesus went to the Temple in Jerusalem with His parents for Mary's purification according to the law (Luke 2:22). It was at this time that the family encountered Simeon and Anna (Luke 2:25–38). He was visited by Magi as a "child" probably some months after His birth (Matthew 2:1–12). In addition, shortly after the Magi's visit the family fled to Egypt to avoid

Herod's wrath (Matthew 2:13–23). Luke tells us, *"And the child grew and became strong; he was filled with wisdom, and the grace of God was upon him"* (Luke 2:40). Finally, we learn:

> Every year his parents went to Jerusalem for the Feast of Passover. When he was twelve years old, they went up to the Feast, according to the custom. After the Feast was over, while his parents were returning home, the boy Jesus stayed behind in Jerusalem...After three days they found him in the temple courts, sitting among the teachers, listening to them and asking them questions. Everyone who heard him was amazed at his understanding and his answers..."Why were you searching for me?" he asked. "Didn't you know I had to be in my Father's house?" But they did not understand what he was saying to them. Then he went down to Nazareth with them and was obedient to them...And Jesus grew in wisdom and stature, and in favor with God and men. (Luke 2:41–52)

Jesus was obedient to his earthly parents. Obedience is the common theme in all three accounts, and this is an attitude that honours our father and mother.

Jacob had his name changed to Israel after he wrestled with God (Genesis 32:22–32). Later Scripture records his son Reuben was disrespectful to his father: *"Israel moved on again and pitched his tent beyond Migdal Eder. While Israel was living in that region, Reuben went in and slept with his father's concubine Bilhah, and Israel heard of it"* (Genesis 35:21–22). An interesting fact concerning the incident is that Scripture makes no other reference to it until the end of Jacob's life. There is no indication of a pregnancy or remorse by Ruben or a reprimand from Jacob, but Jacob did not forget Reuben's disrespect:

> Then Jacob called for his sons and said: "Gather around so I can tell you what will happen to you in days to come. Assemble and listen, sons of Jacob; listen to your father Israel. Ruben, you are my firstborn, my might, the first sign of my strength, excelling in honor,

excelling in power. Turbulent as the waters, you will no longer excel, for you went up onto your father's bed, onto my couch and defiled it." (Genesis 49:1–4)

Reuben's disrespect for his father cost him dearly when Jacob blessed his sons before his death. Reuben's failure to honour his father found him out through his father's blessing years after the sin was committed. Centuries later, during the Exodus, Moses reminded the Gadites and Reubenites: *"you may be sure that your sin will find you out"* (Numbers 32:23). A universal truth to heed!

Moses provides an amazing display of respect for his father-in-law, Jethro, in Exodus 18. First, *"Moses went out to meet his father-in-law and bowed down and kissed him. They greeted each other and then went into the tent"* (Exodus 18:7). The greeting was a demonstration of respect for his father-in-law in that culture, following which Moses shared all that had happened since they last talked. A second act of respect follows:

The next day Moses took his seat to serve as judge for the people, and they stood around him from morning till evening. When his father-in-law saw all that Moses was doing for the people, he said, "What is this you are doing for the people? Why do you alone sit as judge, while all these people stand around you from morning till evening?" (Exodus 18:13–14)

Moses answered him with what today we would call an ego-building "I" statement—*"I decide between the parties"* (Exodus 18:16); as if no one else was capable. Most people like to feel needed, but Moses saw himself as indispensable. In what is now called the Jethro Principle by some church leaders, Jethro replied:

What you are doing is not good. You and these people who come to you will only wear yourselves out. The work is too heavy for you; you cannot handle it alone. Listen now to me and I will give you some advice, and may God be with you. You must be the people's representative before God and bring their disputes to him. Teach them the decrees and laws, and show them the way to live and the duties they are to perform. But select capable men from all the people—men who fear God, trustworthy men who hate dishonest gain—and appoint them as officials over thousands, hundreds, fifties and tens. Have them serve as judges for the people at all times, but have them bring every difficult case to you…If you do this and God so commands, you will be able to stand the strain, and all these people will go home satisfied. (Exodus 18:17–23)

Moses showed Jethro respect when he arrived and greater respect when he *"listened to his father-in-law and did everything he said"* (Exodus 18:24). This act of respect takes on added significance when we realize Moses was now well over eighty years of age, had the wisdom of forty years in Pharaoh's court, and forty years' experience in herding Jethro's sheep. Regardless of age, Moses learned from the wisdom of an older individual. In addition to the Jethro Principle, we can all learn a lesson from the example of how Moses treated the elders in his own life.

You may recall that David has been accused of some serious sin: womanizing and murder, but how did his life end? We already know he took Bathsheba to be his wife and she gave birth to Solomon after the death of their first son. Prior to David's death, however, his son *"Adonijah, whose mother was Haggith,"* (1 Kings 1:5) set himself up as king. Understandably, this was a great concern to Nathan the Prophet who feared for Bathsheba and Solomon's lives. David was not aware Adonijah had taken this liberty, but once informed he said:

"Call in Bathsheba." So she came into the king's presence and stood before him.

The king then took an oath: "As surely as the Lord lives, who has delivered me out of every trouble, I will surely carry out today what I swore to you by the Lord, the God of Israel: Solomon your son shall be king after me, and he will sit on my throne in my place." (1 Kings 1:28 –30)

David then issued instructions for Solomon to *"sit on my throne and reign in my place. I have appointed him ruler over Israel and Judah"* (1 Kings 1:35). The Lord had delivered David "out of every trouble" and we can observe a lot of trouble in his life. However, David would not break a covenant or oath as seen in a covenant made earlier with Jonathan (1 Samuel 20:14–17); and fulfilled through Jonathan's son Mephibosheth (2 Samuel 9:1–13). David had sworn an oath to Bathsheba, saying, *"Solomon your son shall be king after me, and he will sit on my throne"* (1 Kings 1:17). David honoured his commitment to Bathsheba, showing her the respect and love she deserved.

Paul was a man who respected and acknowledged numerous individuals and we see this especially in Romans 16 with his extensive list of personal greetings. I assume some of these would have been older than Paul. One such example is located in his second letter to Timothy:

> *...I constantly remember you in my prayers. Recalling your tears, I long to see you, so that I may be filled with joy. I have been reminded of your sincere faith, which first lived in your grandmother Lois and in your mother Eunice and, I am persuaded, now lives in you also.* (2 Timothy 1:3–5)

These two women, although mentioned nowhere else in Scripture, impressed Paul with a sincere faith and they earned his respect. In fact, Paul respected them to the point of using their life as a means of encouraging Timothy to continue the family tradition. No doubt Lois was older than Paul. He also demonstrates respect when he deals with the Philippian jailer, who may or may not have been an elder to Paul but held authority over him at that time.

> *About midnight Paul and Silas were praying and singing hymns to God, and the other prisoners were listening to them. Suddenly there was such a violent earthquake that the foundations of the prison were shaken. At once all the prison doors flew open, and everybody's chains came loose. The jailer woke up, and when he saw the prison doors open, he drew his sword and was about to kill himself because*

he thought the prisoners had escaped. But Paul shouted, "Don't harm yourself! We are all here!" (Acts 16:25–28)

In Paul's day, for a jailer to loose the prisoners meant his life and so the jailer was about to spare someone else the trouble. However, Paul respected the jailer's authority and his life as a child of God too much to allow him to be slain or to slay himself. The prisoners were free but they did not seize the moment and beat a hasty exit. Paul was a man who showed respect to others in all areas of his life.

Luke must have regarded Theophilus as a man worthy of respect, because he went to considerable length to inform him of the earthly ministry of Jesus. The Gospel of Luke begins:

Many have undertaken to draw up an account of the things that have been fulfilled among us, just as they were handed down to us by those who from the first were eyewitnesses and servants of the word. Therefore, since I myself have carefully investigated everything from the beginning, it seemed good also to me to write an orderly account for you, most excellent Theophilus, so that you may know the certainty of the things you have been taught. (Luke 1:1–4)

Luke, a Gentile and well-educated physician, was writing to either a Roman official or person of wealth as the title "most excellent" suggests. In any case, the respect shown in the letter is very clear. Furthermore, Luke went on to give Theophilus an account of the establishment of the early church:

In my former book, Theophilus, I wrote about all that Jesus began to do and to teach until the day he was taken up to heaven, after giving instructions through the Holy Spirit to the apostles he had chosen. After his suffering, he showed himself to these men and gave many convincing proofs that he was alive. He appeared to them over a period of forty days and spoke about the kingdom of God. (Acts 1:1–3)

What Luke is preparing Theophilus for is the continuation of what Jesus began. By His ministry, "Jesus began to do and to teach" about the kingdom of God. Luke is about to expand on the subsequent events including Christ's ascension, outpouring of the Holy Spirit, and development of the early church. In turn, today's church owes Luke a great deal of respect for these two valuable books in our New Testament.

Obedience and respect tie together the two generations with 'little or no hair, no teeth, eating puréed food, wearing diapers and depending on others for their care.' One discipleship basic is obedience, but this is not possible without respect for the individual seeking obedience. Likewise, discipline, commitment, sacrifice, and weapon handling cannot function without respect. We have interconnectedness taking place as Luke, in obedience to God, recorded two books for Theophilus who he respected. They became available for every generation through the New Testament. Thus, as we obediently read and apply them, we show respect for Luke's effort.

Returning Home
NINE

HEAVENLY HEALING!

Me it is, O Lord, who for mercy does cry out
I am in distress, of this there can be no doubt
My eyes grow weak from sorrow and tears
My soul and body limp from grief's fears
My life by anguish being slowly eaten away
My years groaning by, fading fast as the day
My strength it does faileth from affliction
My bones weak from this deep depression
My foes cause strife as they command
My neighbors in utter contempt do stand
My friends do dread this face to see
Me, when on a street am seen all flee
I am by them forgotten like the dead
I am broken pottery tossed in a shed
I hear many do this good name detest
I am engulfed in terror and denied rest
Me against, it is whom they do conspire
My life to take, their plan of desire.[126]

To God I did quickly call out at last
For the Lord saves us from our past

---

[126] Paraphrase of Psalm 31:9–13

Distressed! Cry out eve, morn or noon
He hears my voice and answers soon
Unharmed by battle, He ransoms me
Though many oppose yet He does see
Yes, God, enthroned forever on high
Hears and afflicts them by and by—

Upon the Lord cast your care,
He will sustain should you dare.
The righteous He'll ne'er let down
On the wicked, God you do frown.
Corruption's pit for deceitful men
Only half their days God will pen.

But as for me, in You I trust.[127]

RETURNING HOME

Returning home from a deployment is a stress-filled event for the entire family, particularly if it was somewhere like Afghanistan. Returning members wonder what has changed during the past few months. How has my household changed? How have my children changed and grown? And most importantly, how has my spouse changed? The rest of the family is wondering how their trooper has changed. A six-month deployment to a war zone can mean significant changes in personality and relationships. Fortunately, often change is minimal and a period of re-adjustment returns life to some sense of normality. Whatever normal is! Collier offers an insightful comment about medals awarded to military members. She writes, "Those medals represent an anxiety in each wife's heart over whether their relationship can ever get back to the way it was. Will it be better or will it be worse? It's certain it won't be exactly the same."[128] After deployment to a theatre of war, with or without metals, life is never "exactly the same." Unfortunately, a percentage of our deployed will return with varying degrees of stress

---

[127] Paraphrase of Psalm 55:16–19, 22, 23
[128] Collier 2004. 97

disorders. The five-phase *spectrum of Combat Trauma*[129] is diagrammed below with Post-Traumatic Stress Disorder (PTSD) at the severe end:

Mild				Severe
I	I	I	I	I
Reintegration Issues	Combat/ Operational Stress Reactions	Adjustment Disorders	Acute Stress Disorder	Post-traumatic Stress Disorder

Not every soldier returns home with PTSD, although an American brochure promoting PTSD treatment suggests in the U.S. it could be as high as one in four. I believe it would be illogical to assume a similar one-in-four PTSD ratio does not exist in our military, as Canadian troops also experience roadside bombings and the death of comrades. PTSD is one issue destined to elevate the heartache and stress level of military family members for long after a physically safe homecoming. Therefore, it is essential that this book draw attention to PTSD, raise awareness of PTSD, and focus attention on the potential impact of PTSD. However, as it is beyond the delimitations of the book to provide anything more than basic knowledge, keep in mind that this primer is not designed to generate experts to work in the field of military stress management or counselling.

What is PTSD? As its name states, it is a stress disorder occurring after a traumatic event or events have been experienced in someone's life. It is not a condition restricted to the military, but can afflict anyone exposed to trauma. Outside of the military, it surfaces most frequently among civilian soldiers responsible for our health and safety. A 'civilian soldier' is my term for the collective occupations of law enforcement officers, emergency response personnel, security guards, prison staff, and any other professionals with a job that brings them in contact with traumatic situations. PTSD may also occur after an individual has experienced a traumatic event caused by severe weather conditions, a

---

[129] Adsit. 22 (Author's emphasis)

vehicle accident, or even simply viewing such scenes. Tragically, every age group is prone to PTSD, even children. Only in recent history has the condition been recognized, named, and studied in adults; little research on its effect upon children has been undertaken to date.

What do we know and what is being done about PTSD? Without question, PTSD is the number one health concern in the military community—internationally. Another concern is Secondary-Traumatic Stress (STS), which can quickly develop into full-blown PTSD. Often STS will impact family members of someone with PTSD. Although in Canada we hear little about either condition, the CF is not uninformed on the topic according to a December 17, 2008 report from Military Interim Ombudsman Mary McFadyen. Her news release states in part:

> While senior military leaders have talked about a strong commitment to deal with post-traumatic stress disorder, or PTSD, and its devastating effects, the commitment hasn't reached down to the community level, interim ombudsman Mary McFadyen said.
>
> There is lack of care and support for soldiers across the country, she said in the 62-page report.[130]

Minister of Defence, Peter MacKay, was quick to respond to McFadyen's claim. Part of his news release issued December 19, 2008 claimed: "One of my priorities is to ensure that our Canadian Forces members receive the best care possible."[131] Probably a number of military family members suffering from STS would like to ask the minister where they fit in his priorities; did he forget or ignore them? January 8, 2009, DND posted a three-page *Backgrounder* news release that included a CF member's loved ones. The opening paragraph stated:

---

[130] http://www.cbc.ca/health/story/2008/12/17/military-stress.html, accessed December 17, 2008.
[131] http://www.forces.gc.ca/site/news-nouvelles/view-news-afficher-nouvelles-eng.asp?id=2831, accessed January 3, 2010.

Military personnel are the most important resource of the Canadian Forces (CF). They and their loved ones deserve the best possible care and support. Over the last ten years, the Department of National Defence (DND) and the CF have put into place a full range of programs and initiatives to contribute to the identification, prevention and treatment of mental health problems such as operational stress injury (OSI), which includes post-traumatic stress disorder (PTSD).[132]

After reading Canadian authors, meeting numerous military families on base, and having discussions with Collier, I am sure there are some families who would take issue with DND's statement: "They and their loved ones deserve the best possible care and support." Some in the military community feel they are not important to DND officials in the least. On this issue, Collier expressed her thoughts to me about DND support for families:

> …when I was researching that book [*My Love, My Life*]…that question came up—"Are you looking at all for support for the wives?" not just for the wives but to see how many might be dealing with PTSD who need a different kind of support—"It's not our mandate!" [she was told]. I know some of the meetings they were having about that time…one woman wanted to know what support is there for my children…they don't know who their father is—"Not in our mandate!" She used some very strong language—[to say] don't tell me it's not in your mandate. They're part of the system too."[133]

The Canadian government started to move in the right direction when they announced the Canadian Forces Mental Health Awareness Campaign on June 25, 2009. In the news release, DND acknowledged the value of our troops, but in the process patted themselves on the back

---

[132] http://www.forces.gc.ca/site/news-nouvelles/view-news-afficher-nouvelles-eng.asp?id=2844, accessed January 3, 2010.
[133] Collier Interview. 15

for their accomplishments. The opening two paragraphs of the release state:

> The Canadian Forces (CF) value their personnel above all else, and embrace the obligation to provide those in need with the very best care and support. Over the last ten years in particular, the CF have expanded and enriched programs aimed at preventing, identifying and treating mental health issues. Additionally, the CF have long sought ways of educating personnel on mental health issues, and creating a culture of acceptance.
>
> There is evidence that these initiatives are working. At a U.S.-Canada Forum on Mental Health and Productivity meeting in November 2008, the CF were praised for their success in reducing the stigma associated with mental illness, which has become a significant workforce and productivity issue in North America in general.[134]

In this press release, as in MacKay's initial release, CF neglects the families of personnel and the issue of STS. Later in the same announcement, there was a comment acknowledging that "military culture is not subject to quick or easy change"—an understatement, considering they admit this has been an ongoing problem over "the last ten years."[135] In addition, as noted earlier by the Military Interim Ombudsman, "While senior military leaders have talked about a strong commitment to deal with post-traumatic stress disorder...There is lack of care and support for soldiers across the country." Christian civilian communities could serve those who serve us by lobbying our government to honour their promise "to ensure that our Canadian Forces members receive the best care possible." At the same time, we should be lobbying for "their loved ones" because DND has acknowledged them as part of their mandate in the January 2009 news release *Backgrounder*.

---

[134] http://www.forces.gc.ca/site/news-nouvelles/view-news-afficher-nouvelles-eng.asp?id=3015, accessed January 3, 2010.
[135] http://www.forces.gc.ca/site/news-nouvelles/view-news-afficher-nouvelles-eng.asp?id=3015, accessed January 3, 2010.

In the United States, PTSD has taken on tremendous importance, especially within the para-church organization Military Ministry, a division of Campus Crusade for Christ International. Military Ministry's concern prompted it to produce a thirty-hour Certificate Training Program known as *Care & Counsel for Combat Trauma*. It was completed in cooperation with the American Association of Christian Counselors and Light University. The brochure promoting the program states, "Over 500,000 American veterans now suffer with untreated PTSD." In addition, it claims: "one in four veterans who return home from combat suffers with the effects of *Post-Traumatic Stress and Combat Trauma…*"[136] It seems reasonable to assume that similar unfortunate ratios exist among our veterans of war zone deployments. I have completed this training program as part of my research for the book; it is excellent, but in my opinion beyond the requirements of the average civilian. However, segments of the program helped me to understand PTSD, which makes the series valuable as civilians strive to support our troops by understanding the lifestyle. PTSD has become a component of this lifestyle and is destined to remain part of it for many years to come. Those who want to support our troops need to be aware of PTSD and its potential consequences, but it would be impractical to think average civilians could develop stress management or counselling skills to assist PTSD victims.

What are the basics concerning PTSD? It would be unreasonable to ask people to read extensively about PTSD, but as mentioned earlier, selected sections in the video series are helpful. It would be beneficial for civilians to be aware of basic PTSD statistics and terms such as trauma, crisis, stress, and PTSD. Collier contributes basic information with the following concise and helpful comments:

> As one military wife said to me, *Please explain PTSD to me so that I can understand it.* According to the DND website, *Post-traumatic stress disorder is a psychological injury caused by the*

---

[136] Military Ministry Publication. Bridges To Healing Series. *Care and Counsel for Combat Trauma: Professional Certificate Training Program*. n.d. A four page brochure, unnumbered. (Author's emphasis)

reaction of the brain to a very severe psychological stress such as feeling one's life is threatened. A professional explained it to me this way: *A critical incident is a traumatic event that goes beyond our capabilities to cope. It is a psychological wound. Post-traumatic stress refers to the symptoms one may develop after a critical incident, such as increased heart rate, nightmares, flashbacks, thoughts of suicide, etc. Post-traumatic stress disorder results when a person cannot cope and the symptoms persist over a period of time causing problems in a person's ability to function.*[137]

Collier's comments are an excellent starting point for civilians to develop a framework to understand PTSD. This can be simplified to Adsit's statement noted earlier: "Post-traumatic Stress Disorder is a *normal* reaction to an *abnormal* event"[138] although there is nothing simple about PTSD. In his research, Grossman draws attention to the "Relationship between Degree of Trauma and Degree of Social Support in PTSD Causation." He presents this as an equation:

"Degree of Trauma $\times$ Social Support = Magnitude of Post-Traumatic Stress."[139]

Floyd basically concurs with Grossman when he writes:

A key factor that mitigates how a person responds to trauma is the individual's level of social support before, during, and following a traumatic occurrence. A person who has close, supportive relationships has a better likelihood of coping effectively.[140]

I believe this is key information for Christian civilian communities to keep in mind, because it points to the only area in which we can offer

---

[137] Collier, 2004. 169 (Author's emphasis)
[138] Adsit. 23 (Author's emphasis)
[139] Grossman. 283
[140] Scott Floyd. *Crisis Counseling: A Guide for Pastors and Professionals.* Grand Rapids, MI: Kregel Publishing, 2008. 60

significant assistance—social support. Grossman and Floyd indicate high support levels reduce the incidents of PTSD even in high-trauma situations. In other words, they maintain support makes a difference in the lives of our troops. The PTSD problem is part of our military today and these two experts are saying support is part of the solution. *This is about as basic as it gets.* Where better for support to begin than in the communities of Christ's followers?

Biblical Parallels

Only one biblical personality provides any hint of dealing with Post-Traumatic Stress Disorder. The individual is King David, a man whose life I have drawn upon repeatedly before, although for this subject Adsit explains:

> Not only was he a man of great courage, a brilliant military leader, and the most powerful king the nation of Israel ever had, he was also "a man after God's own heart" (Acts 13:22). But David was one other thing that surprises a lot of people—though it shouldn't. David also was a PTSD sufferer. How could one think otherwise when you read passages like this which David wrote:

> *Psalm 31:9–13 Be gracious to me, O Lord, for I am in distress; my eye is wasted away from grief, my soul and my body also. For my life is spent with sorrow, and my years with sighing; my strength has failed because of my iniquity, and my body has wasted away. Because of all my adversaries, I have become a reproach, especially to my neighbors, and an object of dread to my acquaintances; those who see me in the street flee from me. I am forgotten as a dead man, out of mind, I am like a broken vessel. For I have heard the slander of many, terror is on every side; while they took counsel together against me, they schemed to take away my life.*[141]

---

[141] Adsit. 165—Adsit quoted Psalm 31 from the New American Standard Bible (emphasis added)

For this parallel, I am indebted to Adsit for drawing attention to the PTSD aspect of King David's life in Appendix D of *The Combat Trauma Healing Manual*. Adsit continues:

> David wrote over seventy-five desolate, anguish-filled passages like this in the Psalms…David recognized God was his Healer… When you read many of David's Psalms, you are reading the writings of a man in progress. He fought with depression, guilt, fear, anger, despair—probably many of the same emotions you fight with.[142]

In other words, David experienced many of the same emotions CF members and their families deal with on a daily basis. In most of his Psalms, David turns to God; he chooses to look up, not down, as he does in Psalm 31:14–22. Adsit provides six pages of Psalm passages and commentary relative to PTSD suffering. Adsit currently has two manuals available—one for the returning warrior and one for his spouse (specifically designed for women) with additional manuals scheduled for release. The Bibliography lists the two available manuals. This series would make a valuable resource for a church library, essential for those interested in a serious commitment to military ministry.

Our examination of lifestyles issues is complete. What remains is an inquiry into the work of a very special and valued group in the CF. At His crucifixion, Jesus Christ represented both the High Priest offering sacrifice as well as being the sacrifice for the sin. A chaplain offers ministry to military lives while at the same time living life in the military. The poem opening the next chapter looks at passages from Deuteronomy and Joshua to review the origin of chaplaincy.

---

[142] Adsit. 165

Chaplain's Corner
TEN

CHAPLAINCY GENESIS

War is the result of sin, this we all know.
Before going to war we should act so slow.
God knew our sinful desire would dominate,
So provided away on Him to contemplate.
In the Word God said:
When against your enemy in war you do plan;
Seeing horses, chariots, and army to terrify man.
Fear not!
The Lord God said:
For the Lord your God out of Egypt brought you,
Never I'll renounce, forsake, or forget about you.
Fear not!
Then the Almighty said:
When to a battle you are about to enter in,
My priest will speak first to forgive your sin.
Yes, he shall address the army band.
First, ahead of all those in command.
Fear not!
The priest will say:
Israel, today against enemies you will battle;
Weakness, fear, terror or panic you can't rattle.
Fear not! Fear not!
The One going with you is the Lord your God,

GOD'S TACTICAL UNIT

> To fight for you against the enemy and his god.
> The victory to secure![143]

> Jericho's fall was God's long-term practical plan;
> In the land, Joshua's spies to Rahab's house ran.
> She and her family all would be spared,
> For their secret with others was not shared.
> Take up the Ark of the Covenant was Joshua's order
> With seven priests ahead carrying trumpets to thunder.
> God personified, His earlier command now in action
> With the Ark and the priests to His satisfaction.
> Round the city to march when Joshua said: Advance!
> Armed guard took the lead with Ark next in stance;
> The armed guard marched off to head the invasion
> With priest's blowing trumpets to mark the occasion.
> And the rear guard followed at the procession's closing;
> As the trumpet sound blasted to send enemy fear rising.
> The victory to secure![144]

CHAPLAIN'S CORNER

Whether they are known as chaplains or padres, the work of those in chaplaincy is significant in military life. When did such a role begin? I trace it back at least as far as Moses because of this instruction:

> *When you go to war against your enemies and see horses and chariots and an army greater than yours, do not be afraid of them, because the Lord your God, who brought you up out of Egypt, will be with you. When you are about to go into battle, the priest shall come forward and address the army. He shall say: "Hear, O Israel, today you are going into battle against your enemies. Do not be fainthearted or afraid; do not be terrified or give way to panic before them. For the Lord your God is the one who goes with you to fight for you against your enemies to give you victory."* (Deuteronomy 20:1–4)

---

[143] Paraphrase of Deuteronomy 20:1–4
[144] Paraphrase of Joshua 6: 6, 7 & 9

CHAPLAIN'S CORNER

The priest in this case is exceedingly important, as he speaks first to the army. This is similar to the advance against Jericho when orders from the *"commander of the army of the Lord"* (Joshua 5:14) were given to Joshua. The strategy called for the Israelites to march around the city walls; subsequently Joshua issued this order:

> *"Take up the ark of the covenant of the Lord and have seven priests carry the trumpets in front of it." And he ordered the people, "Advance! March around the city, with the armed guard going ahead of the ark of the Lord." ... The armed guard marched ahead of the priests who blew the trumpets, and the rear guard followed the ark. All this time the trumpets were sounding.* (Joshua 6:6–7, 9)

As I interpret the passage, in the battle and victory over Jericho, God, personified by the Ark and priests, served as chaplain to His people.

If we look slightly beyond the New Testament, Cole offers this interesting insight on Eusebius of Caesarea (A.D. 260–339)[145]; this also reminds us of a comment by Origen (A.D. 185–254) I presented in chapter one:

> Eusebius saw the ordinary Christian as being able to achieve some sort of Christian perfection while remaining in an earthly station, including a military station. Such a perfection was seen as secondary to the spiritual elite, but it was a real perfection nonetheless. Meanwhile the spiritual elite were to do the most good not fighting for the emperor but by praying for him. There's even evidence that Eusebius was the first to propose a kind of military chaplaincy.[146]

In any case, the starting point of chaplaincy may be debatable but it certainly has roots deep in our Judeo-Christian heritage.

If we press the fast-forward button, we can find an early mention of Canadian chaplaincy some two hundred years ago. In Chapter two, I

---

[145] Cole. 12
[146] Cole.14

shared that "some priests had served as chaplains in the army"[147] during the War of 1812. If we jump forward another one hundred years, we learn that "Chaplains were recruited for overseas service and pastoral duties were enlarged to cope with the trauma of war at home"[148]; in this case, the First World War. Chaplaincy has a long history in this country, and it continues today as noted in chapter two with this quote from the Interfaith Committee of Canadian Military Chaplaincy:

> Historically, the Chaplain Service in the Canadian Forces has been rendered by a partnership between the Government of Canada, as represented by the Minister of National Defence, and the Canadian Conference of Catholic Bishops and the Canadian Council of Churches. With the evolution of the Service the partnership included churches and faiths beyond the foundational agreements, in order to reflect the diversity of Canadian Society and to meet the needs of all in the Canadian Forces.[149]

It became apparent while researching this project that chaplains maintain a low profile. A modest group, they are seldom mentioned in civilian conversations, rarely cited in the newspaper or on television, and infrequently write or are written about in books. Chaplains truly are unsung heroes in our military. The objective of this chapter is to focus on the chaplain's role in the daily and spiritual lives of our troops and their families. However, it should be noted that we have a responsibility to specifically remember the chaplains in terms of support and prayer. After all, they are a unique branch of our CF, deserving support and prayer along with other men and women who wear a Canadian military uniform. One reason chaplains may be neglected can be found in a thought Cash expressed: "There are some places where you just don't expect to find God." Which begs the questions: is the military one of

---

[147] Murphy. 88
[148] Rawlyk. 143
[149] http://www.cmp-cpm.forces.gc.ca/cfcb-bsafc/pub/iccmch-gciamc-eng.asp, accessed February 21, 2013.

those places where people don't expect to find God? If so, why? Should we not expect to meet God anywhere and everywhere in His universe? Is God not depicted as a warrior in both the Old and New Testaments? He fought temporal evil in the Old Testament; He battled spiritually in the New Testament and He continues to battle evil in both realms even today. Cash, in his comment, was referring to Baghdad. He adds, "This was no place for God; this was enemy territory. But when the smoke settled, God's footprints were clearly visible."[150] God was there! We can count on Him to never leave or forsake us even in downtown Baghdad. Cash made this discovery during a U.S.-led invasion of Baghdad in March 2003.

In this chapter, I will draw information from books by chaplains as well as my own experience in working with and discussing chaplaincy issues with chaplains at CFB Borden, CFB Petawawa, and MCF annual meetings. The three books I will quote are by Coleman and Davis, Canadian authors, and Cash, an American. In the chaplain tradition of modesty, Davis and Cash dedicated their accounts of serving God in the military to the men to whom they ministered. Both authors ministered exclusively to males in the situations they shared. Coleman offers an historic novel based on his experience as a chaplain. The Canadians wrote about World War II, and as mentioned Cash's work is more current. The chaplain's role in the 1940s and now is noticeably similar; it seems some things never change.

As a civilian writing about military chaplaincy, I am not able to share experiences exclusive to their situation. I have served as a pastor across denominational lines, and this provides a limited resemblance. However, the nature and condition of chaplaincy varies too much for anyone except a chaplain to be able to relate effectively. Therefore, the approach taken here is to allow chaplains who have written about their experiences to share them. As a result, I will use numerous quotes from these three books in an attempt to present chaplaincy ministry by outlining what tasks they find in real life. Civilians may not be aware

---

[150] Lt. Carey H. Cash. *A Table In The Presence: The Dramatic Account of How a U.S. Marine Battalion Experienced God's Presence amidst the Chaos of The War In Iraq.* Nashville: Thomas Nelson, 2004. b.c.

of all the roles chaplains are called upon to fulfill, so three of the most difficult are examined here.

The first task is and always will be difficult for chaplains: the burial of a lost comrade. Davis writes about two funerals:

> I thought of another body I'd buried. It was my first experience conducting a burial for a casualty of enemy action—a boy who died, near where we were stationed, from wounds received at Dieppe...
>
> At the end of the service...thoughts went across the seas... where a family would be bravely trying to cope with the overwhelming loss of a son, a brother, a husband. I had to pinch my thigh to get control and give the final blessing. As I looked into the eyes of the burial party, I knew I was not the only one who had to struggle for control.[151]

Ralph, the chaplain in Coleman's novel, shares three death incidents in the "final engagement before the fall of Sicily."[152]

> [Major Jock]MacPherson moaned as Ralph gave him the rites that he had requested. It was not the first time he had performed this act. He and the only Roman Catholic chaplain in the brigade, Father Barker, had agreed to do this for each other, when one of them was absent from the scene of battle. [First]
>
> "Jesus, Jesus, Jesus...help me. Oh, oh, oh..."
>
> "I know son, give me your hand. Now, hang on while we pray. You don't have to say anything. Just let's us be together with the Good Lord."
>
> The two men held hands together for some minutes until Ralph felt the boy's hand go slack. The chaplain looked into the glazed eyes and drew the lids down.

---

[151] Davis. 39
[152] Coleman. 169

CHAPLAIN'S CORNER

Ralph's throat constricted. "Sleep well, Jim. You fought the good fight. I'll tell your folks I was with you right to the end." [Second]

Later, when the battle line moved on...Ralph led the burial party to the gully where B Company's dead lay. Each body was wrapped in a gas cape. The ones with missing body parts were placed to the side, while attempts were made to match pieces with owners. It was such a ghastly task. Shallow graves were dug, and bodies were lowered into the dusty, etched-out ground. Ralph offered prayers and a piper played... [Third][153]

Today, military procedure is to repatriate a fallen member of the military for burial. However, Cash does share his reaction to a Killed-in-Action (KIA):

"Casualty! Casualty! Two Marines down at Pumping Station 2. Request chaplain."

Rufo and I ran back to our vehicle and waited for the security team to come for us and to escort us to the site.

Within minutes the dust of a speeding Humvee signaled the approach of Captain Dickens, whose vehicle was mounted with a Mark-19 grenade launcher on top...The landscape was just too dangerous for Rufo and me to drive alone, especially in a non-armored vehicle like ours.

...we finally reached our destination...

"KIA. One of them was killed in action."

My heart sank. My impulse was to grab Dickens and shout, "No! It's not true. Nobody's dead!"

Instead I quietly asked, "Dear God, who was it?"

"It was Shane."

In the hours that followed, as I talked to the men who had been with him, I was able to piece together the last hours of Lieutenant Shane Childers's life."[154]

---

[153] Coleman. 170, 172/3, 173
[154] Cash. 77–9

It was not just the death of Lieutenant Childers that Cash had to deal with. He also spent hours talking to the other men who witnessed the incident. Survivor guilt is real and it does impact traumatic stress levels. Early support intervention is important, and here the chaplain on the scene initiated a support process, thus accomplishing a great deal to alleviate the traumatic effects.

Grief counselling is not the only counsel a chaplain is called to provide. As mentioned earlier, military divorce rates exceed national rates; it was and continues to be a problem. It is not uncommon for women and men in the field or on the home front to consult with their padre on this and other issues. As Davis notes, "At least fifty percent of the problems, presented by men seeking advice and help, sprang from marital relations or non-marital liaisons."[155]

Probably the second-worst job is informing family members of their loss, but chaplains are often rewarded for their dedication to this responsibility. Davis sites this example:

> One evening a young gunner came and told me his fears. He said, "Padre, if anything happens to me it will kill my mother; Father went to France in the First World War, before I was born. I never saw him, and he never saw me. What a rotten break it would be if she lost me too. In spite of that, I don't want to leave the regiment, but I hope for her sake I make it back."
>
> The next day, one shell landed in our lines. The young gunner was in a tent typing a letter. One piece of the shell splinter pieced the wall of the tent and hit him in the spine. He was dead before we got to him.
>
> I wrote ten letters to his mother and tore each one up. Finally I wrote one, signed and sealed it, and sent it off. Some weeks later I got a reply, one of the bravest letters I ever received. She said that she had been richly blessed by having gained the love of two of the finest men that ever lived. She said, "You knew

---

[155] Davis. 24

CHAPLAIN'S CORNER

my son well. His father was just like him. I know many women who have lived a whole lifetime and never known what I have known, and have never been blessed as I have been blessed."

This lady added to my conviction that women have a great and wondrous inner strength and courage.[156]

This must have been an encouraging letter for Davis to receive, but it also says a great deal about the women of our nation and their contribution to the war effort. Women like this one, serving on the home front, fortify our front line troops. They play a very significant role in winning any war! In Ralph's account, quoted earlier concerning Private Jim Johnson, their pre-death conversation is recorded below; one can imagine the difficulty this padre had composing a letter to his parents. "The boy was from Plevna,[157] and Ralph knew his family well."[158]

"How are you feeling, Jimmy?" asked Ralph, wiping the sweat from the soldier's eyes.

"Terrible! Can't feel my legs! Cramps in my stomach!"

Ralph looked bleak. "I'll get some more morphine for you."

"No, don't leave me, Padre," said Johnson weakly, tying to rise up. "I think I'm a goner."

The chaplain gently laid the soldier back. "Just take it easy, son. Is there anything I can do for you?"

"Will you write to my mom? Will you tell her what happened?

Tell her I miss her—and Dad."

"That I will, Jimmy."

"I wish I had been a better son. I gave them a pretty hard time when I was in high school." The boy coughed up blood

---

[155] Davis. 122
[157] Plevna is a small, rural community in the Addington Highlands approximately 150 km. or 2 hours west of Ottawa, ON.
[158] Coleman. 172

and whispered. "Didn't study...got in with a bad bunch... raised hell!"...

Ralph bent closer. "They knew you had a good heart, Jim. You just had to get some things out of your system."

"Oh, God, I hurt," wheezed the soldier.

Ralph continued. "I remember talking with them just before I joined the regiment. You had already signed up. They said, Look after Jimmy for us. He may be a little wild, but we love him. Deep down, he's a good boy."

"Jesus, Jesus, Jesus...help me. Oh, oh, oh..."[159]

Cash gives no indication of writing to families, because death notifications are now handled by chaplains on the home front. However, he does provide a significant and reassuring thought in this reflection on the life of Childers, marines, and Christianity:

In the infantry, when a man falls in battle, the most senior man beneath him must step up and assume the place of the fallen leader. That man was Staff Sergeant Bradley Nerad...I had just baptized Nerad as a new Christian about two weeks earlier... that had inspired more than half of his platoon—Shane's platoon—to follow.[160]

...I stood staring at the saddened faces of Alpha Company after they'd lost their lieutenant. I knew they loved him. I loved him too. Yet despite the sadness I felt, I couldn't help but reflect upon the providence of God in it all. From the beginning, He had known this hour was coming—a dark and difficult hour, an hour of almost unbearable sadness for some of us. Yet as a Good Shepherd cares for His sheep, God had provided for these men everything they would need to sustain and strengthen them.

God had given them a company commander who saw it as one of his highest priorities to ensure their spiritual lives were attended to.

---

[159] Coleman. 172
[160] Cash. 88

God had given them a first sergeant whose winsome life was a living embodiment of Jesus.

And God had given them a tough staff sergeant whose public step of faith had shown them that a U.S. Marine can be a rugged man and warrior and still follow Christ with the faith of a child...He was with us that dark day on the morning of March 21.

Gathering the men together around the AAV where their lieutenant had been lying just hours earlier, we observed a brief service of committal and entrusted the soul of our fallen brother to the all-merciful hands of our Father in heaven. And we knew we were not alone.[161]

If Cash did write the parents of Lieutenant Shane Childers, he had both good news and bad news to share with them. In another incident, Cash recalls what happened when the troops were ordered to move north toward Baghdad on March 17, 2003:

A young officer grabbed my arm. "Hey Chaplain, if something should happen to me, will you give this letter to my girlfriend?"

Hey, wait a minute, I thought. *That's just for the movies.*

The man who handed me the letter was a decorated combat veteran. He had been among the first to land deep in enemy territory in Afghanistan. He'd been through this before, and his face was quite serious. I took the letter.

"It's going to be OK," I reassured him.

Was I sure about that? I tucked his letter deep within my pack, quietly hoping that I would never have to pull it out.[162]

This is an interesting exchange between a combat veteran preparing for the worst and a chaplain who doubts his reassurance to this marine. However, chaplains are human and subject to the same doubts and fears as each of us.

---

[161] Cash. 93–4
[162] Cash. 2 (author's emphasis)

The final role to draw to my reader's attention is the one of chaplain as a soldier. Yes, they serve as soldiers for Christ Jesus; but in addition, a chaplain is a paid member of the military. Cash draws attention to a chaplain's status:

> "Sir, do you need any help loading your supplies?"
>
> It was a voice I recognized well, even in the dark. Second Class Petty Officer Redor Rufo was my personal assistant—a religious program specialist (RP), who assists me in the administrative tasks required to conduct ministry. That is not, however, his only job. In war, the RP is a chaplain's bodyguard. Geneva Convention and navy regulations do not permit chaplains to serve as combatants. In fact, a chaplain is the only member of the entire military who is not permitted to brandish a weapon. Chaplains sometimes speculate about what they would do if their men were overrun, if they had the choice to defend themselves with a rifle or pistol. It's a tough call, but I always thought that if the life of a fellow Marine or sailor was at stake, I'd use the weapon.
>
> Nevertheless, I knew Rufo would not hesitate to use his rifle if we ever found ourselves in real danger.[163]

This descriptive account of chaplaincy policy points out the vulnerability of chaplains, yet at the same time demonstrating a concern for their safety. It also divulges a quandary chaplains face regarding personal self-defence. Considering the office represented, however, it would be difficult to respect a weapon-brandishing chaplain offering services on behalf of Jesus Christ, the Prince of Peace. Although as noted below, one can only imagine there are days when even the chaplain would like to be armed.

> "Get down! Get down! Don't know where it's coming from, but get down...now!"

---

[163] Cash. 6

Second Lieutenant Joshua Glover's urgent plea didn't fall on deaf ears, as every one of us either hit the deck or grabbed a weapon to fire back.

Stunned by the wicked melee of machine-gun fire, I lunged over and threw myself on the ground behind the front bumper of my Humvee. Rufo was right behind me. With his trembling hand on my shoulder to keep me down, he immediately made ready his rifle and prepared to engage. A hail of bullets was raining down around us. Where was it coming from? We had no idea, but it was close, way too close.

It was the morning of April 9. I had just been leading a worship service for a platoon of Charlie Company Marines on the banks of one of the lush Tigris River tributaries.[164]

April 9 was the day before the Marines launched an attack on and entered Baghdad, and during the attack Chaplain Cash was not far from the action, as he states in the following account:

Assuming that a well-armed and determined enemy force would be waiting for us at the palace, Lieutenant Colonel Padilla did something that up until now he had not done before. He ordered all of our non-armored vehicles (mainly the soft-skinned Humvees and seven-ton trucks that comprised the combat train of which I was a part) to proceed with the convoy only as far as the outskirts of the downtown Baghdad area. There they would await further orders.[165]

The fifteen vehicles in our combat train pulled off the side of the road to let the armored convoy pass. We watched them disappear, one after another, down Route 2's shadowy four-lane highway.[166]

---

[164] Cash. 157
[165] Cash. 176–7
[166] Cash. 179

For me this is the second of four illustrations pointing out that some things never change. Cash, a chaplain in the twenty-first century, is with his men going into battle. Is this not parallel to the presence of the priests as Joshua marched on Jericho? In the two following accounts, those on the front line also walk with God's representative in their midst.

The story was no different in World War II when Chaplain Davis accompanied the troops on their entry into Sicily. This dramatic account places him right at the heart of the battle:

> …we landed on the southeast beach of Sicily…
> That night I lay in a slit trench on the beach and had a very good view of German and Italian bombers flying overhead. We could see the bombs fall from the planes on one side of us, and travel over to land on the other side. I had just rolled over on my stomach when a large piece of flesh fell into my trench and hit me on the back of my neck. At first I thought that I had been hit, and the flesh was my own. It was with relief I verified I was intact and the flesh was not mine. It was a disgusting but happy realization. However, the episode ended my attempts to eat supper.
> In the morning I found that a shell-shocked cow had been torn apart by a bomb, and part of her had fallen into my hideout. Perhaps it is hard to believe, but I was grateful when I learned I had been hit by a three-pound raw chunk of bloody beef.[167]

The main character in the historical novel by Coleman was not as fortunate as Cash and Davis. However, his injuries may well be factual since the story is based on true events. It also represents a fourth example of God's central role upon the battlefield:

> It was a Sunday morning and Ralph was delivering the sacraments to D Company's location. He would meet with little groups of four or five men at a time, read a couple of prayers and distribute the elements. As he approached the platoon HQ [headquarters]

---

[167] Davis. 42

position, the air suddenly filled with the sound of incoming *nebelwerfers*, a type of mortar used by the Germans and called "Moaning Minnies" by the Allied troops. The explosive charge was not only deadly, but psychologically unnerving.

As Ralph was pulled to cover into the platoon commander's slit trench, the ground erupting behind him, a blast of shrapnel pierced his left arm and back. He was thrown face down into the dugout, and blood soaked through his ripped tunic. The force of the blast stunned the other men in the hole, and it took a moment for them to assess the situation. The company MA [Medical Assistant] made a quick evaluation and called for an ambulance to come forward to evacuate the chaplain. Lieutenant Marty Kossowski retrieved the communion kit Ralph carried in a small pack and placed it on the stretcher between his feet. Then, the stretcher-bearers carried Ralph to the waiting vehicle.

"How is he?" asked Corporal Mickey Maguire.

"Unconscious, but still breathing. Looks pretty bad, though!"...[168]

"There'll be some cheering...when they get the news he's going to be fine...They hold their padre in high esteem. He was clearly a soldier's padre or, I should say, is."

"Padre, you say? Well the Good Lord was surely with him this time..."[169]

Ralph sighed. "I'm going back to England, but what about my lads? Can't they patch me up here?"

"Ah sure an' they can't here. The reconstruction will require a specialist's touch and rehabilitation..."[170]

Notice Ralph's first thought—it was his men! Ralph's experience is not unusual, as serious injury and death are not uncommon for chaplains in war.

---

[168] Coleman. 209
[169] Coleman. 213
[170] Coleman. 215

Unfortunately, chaplains are not always present nor ordered to the front line as the following comments by Hope indicate:

> In the darkness and dust, amidst the noise and confusion, this group of Canadians came under intense small arms fire, wounding three and killing Pte Robert Costall...
> It became apparent that our casualties were possibly a result of friendly force fire from US/ANA [Afghanistan National Army] position ....a US Army Colonel and a chaplain (...went to... administer pastoral support to US troops and to commemorate the loss of one of their soldiers) ... Our soldiers watched as the US Colonel and Chaplain provided leadership and moral support to their troops while the Canadian representative informed these tired and hurting troops that their casualties were possible from a blue-on-blue and that they had to cooperate with him. They were furious at this insensitivity and lack of leadership.[171]

This incident is included not to disparage Canadian chaplains or the CF but to indicate that each battle situation is unique and chaplains are not always present or available. In fact, Hope later discovered "that a formal investigation into the incident commenced that day"[172]—the day following the casualties.

Chaplains wear a number of hats and I have looked at three of probably the most difficult ones: a funeral director burying his comrades, a long-distance comforter to the home front, and an unarmed soldier. Chaplaincy is an emotionally, mentally, physically, and spiritually draining responsibility. Who ministers to the chaplain or padre in battle? I believe chaplains need to develop relationships before deployment with other chaplains and pastors in civilian ministry. Local pastors need to understand the chaplain's lifestyle and seek to communicate with them. The first step pastors might consider is extending an invitation to a base chaplain to speak to their congregation or encourage them to attend local ministerial meetings. This would allow a pastor to discover what

---

[171] Hope. 72–3
[172] Hope. 73

ministry can be offered from the home front in addition to prayer while a chaplain is serving on the front line.

In the field, and at great risk to their personal safety, chaplains have served as an interdenominational shepherd, funeral director, mortician joining body parts, assistant medical officer, grief counsellor, comforter to the dying, communicator to the home front, and more. When no longer in a theatre of war but safely home on base, their assignments may lessen somewhat. Still, the minimum for a chaplain includes trans-denominational shepherding, counseling, comforting, and communicating. In their role as a communicator, death remains a difficult message to deliver. This point is clearly made in the Casualty Administration Manual, Chapter 1—General Information. Note the acronym PEN stands for Personal Emergency Notification.

Step 2—Notification of the PEN contact
32. The person identified in section 3 of the CF 742 (PEN form) must be notified of the death in the most timely, respectful and dignified manner possible. Ideally, the CO should conduct the notification in person. If not practical, the CO or his/her superior commander may designate an officer of the appropriate senior rank to conduct the notification. The unit chaplain, duty chaplain, or, if no military chaplain is readily available, a civilian clergy-person preferably of the same faith as the casualty, should accompany the officer performing the notification. The officer performing the notification, in consultation with the PEN Contact, will determine which other family members should be notified, whether CF assistance is required, and advise the chain of command when the process is completed. Most importantly, after notification, the CO and the chaplain will ensure that immediate support is offered to the NOK [Next of Kin].[173]

Hopefully this helps explain some of the burdens placed on chaplains as they live and work within the military lifestyle. Chaplains

---

[173] http://www.cmp-cpm.forces.gc.ca/cen/pub/cam-map/cha-01-eng.asp, accessed February 21, 2013 (see "Commanding Officers Guide").

and the CP provide an invaluable asset for our troops, and as such they must not be forgotten within military ministry. The average Christian civilian may not be equipped to minister directly to a padre in his/her battles, but church pastors can in some ways be considered part of a padre's peer group. Support for chaplains is one area of the slogan 'we support our troops' that rests solely with local ministers and pastors. It is, however, the responsibility of all Christians to pray for our troops and their chaplains.

Biblical Parallels

In Deuteronomy, we read about God's call for a priest to address the army prior to the CO. Next, we recognized God's presences through the Ark of the Covenant and priests at the Battle of Jericho. These two were mentioned in this chapter and represent God's interest in chaplaincy ministry. The Israelite battle against the Amalekites with Moses standing *"on top of the hill with the staff of God"* in his hand (Exodus 17:9) is parallel to either home front support or the Jericho fight. The Amalekites were engaged prior to construction of the Ark. It was a time when Moses was God's official spokesperson to the nation—in many ways their chaplain. As well, on the hilltop Moses stood between the front line and home front. God does not maintain set patterns for handling individuals or situations, as I am sure most readers have discovered. In fact, after Jericho, He frequently spoke directly with CO Joshua and the Judges even to the point of issuing battle strategies. God's power cannot be used frivolously, however, as the Israelites attempted to do in a battle with the Philistines. When Israel lost four thousand men, they sent for the Ark of the Covenant and because of their rejoicing upon its arrival *"the Philistines were afraid"* (1 Samuel 4:7) but still fought. The Philistines defeated the Israelites, killing *"thirty thousand foot soldiers"* (1 Samuel 4:10) and the *"ark of God was captured"* (1 Samuel 4:11). This is an example of the Israelites placing their confidence in the ark (it became an idol) instead of God who earlier had instructed them: *"You shall not make for yourself an idol in the form of anything in heaven above or on the earth beneath or in the waters below"* (Exodus 20:4). God's presence is not dependent upon the Ark of the Covenant or military chaplains acting on

His behalf. Neither the Ark nor a chaplain can guarantee victory; God alone should be sought, trusted, and worshipped.

In the New Testament, there are three possible parallels; I say 'possible' because using them may stretch the chaplaincy parallelism to the limit. Thus, I rolled these three examples into one question. Are Philip's encounter with the Ethiopian (Acts 8:26–40), Peter's visit to the home of Cornelius (Acts 10:1–48) and Paul's experience aboard the ship to Rome (Acts 27:1–28:16) examples of chaplaincy ministry? These situations involved God's person positioned for ministry in situations similar to where a chaplain might be found. Nonetheless, the three represent unique, unusual encounters, making me hesitant to call them examples of chaplaincy ministry. This is, however, an interesting thought to ponder as we close the chapter; I leave my readers to individually decide the chaplaincy significance of these New Testament possibilities.

Cease-fire!
ELEVEN

CEASE-FIRE!
As I understand a cease-fire, it is usually called for one of two reasons—either one of the sides has waved the white flag of surrender or both sides simply agree to stop firing at each other because further action would result in unnecessary carnage. For our purposes, this book must eventually come to a close—however, the spiritual war surrounding us is far from over. This is but a temporary cease-fire.

> Cease-fire! All shooting must now terminate;
> Here only used to signal my book's closing state.
> Certainly Satan, the war has not finally won—
> His fury continues until the coming of the Son.
> No! To my enemy I have not conceded the fight,
> Though many a battle has raged, day and night.
> A white flag in surrender will never be waved,
> For on the Cross Jesus died so all may be saved.
> This war has waged since books' conception stage;
> Months—no years—long ago before the first page.
> But now is a time to review what's been done,
> And in the process created this poem just for fun.
> Where have we been exploring the military scene?
> A primer it's called, thus accordingly quite lean.
> Hope it raised awareness of our troops to support,
> Serving home, abroad, and in places of all sort.

> Hope it proved a call to serve Christ their way,
> To sacrificially obey His commands day by day.
> Called in faith—to trust—have hope—give love
> We use our weapon to seek guidance from above.
> Our call is love neighbour through God's grace
> As we share with them on this earthly space.
> Now in turning to thoughts with which to close;
> Trust these give Satan a final punch in the nose.

CEASE-FIRE!

This book set out to prepare readers for a fully active life in God's tactical unit and to raise awareness of the need to support our troops in the process. Opening chapters studied 'priority' information and history arguing churches could learn *about* military life and by doing so open a door allowing them to learn *from* the military in their midst. 'In their midst' is not limited to churches in close proximity to a CFB, but includes any church that might deal with Reserve forces, veterans, and family members spread across Canada. My argument presents members of our military as disciples of the CF. They are trained to develop:

1) Discipline to build faith and trust in their commanding officer
2) Commitment to serve their country
3) Obedience to orders and commands
4) Willingness to pay the supreme sacrifice if necessary
5) Trustworthiness in weapon handling.

Are not these also the primary goals for Jesus Christ's disciples? When called by Christ, we are to be disciplined, committed, obedient, sacrificial, and trustworthy in handling God's Word. In Christian and military training, the discipline of faith is essential, for *"without faith it is impossible to please God [or one's commanding officer]"* (Hebrews 11:6). Both Christian and temporal soldiers must have faith in the authority in charge and faith in the mission. Faith is fundamental to serving a cause, whether it is spreading the Gospel or defending the nation. We are the people of faith, yet we also constantly need to strengthen our faith and share it with others. Boot camp represents the starting point for military soldiers and should be the commencing point for *"a good soldier of Christ Jesus"* (2 Timothy 2:3). Temporal soldiers do not simply fill a uniform and play war games, nor should Christians dress up and only warm a pew. Soldiering requires a great deal more!

We graduated from boot camp to study the emotional issues encountered when war is declared, a time of living with war or its threat. This chapter raised awareness of the emotional upset endured by our troops and their families, demonstrating the sacrifices they make on our behalf. The lifestyle entails living with an uncertain tomorrow, wondering if loved ones will return from deployment, being uprooted because of war, and experiencing discomfort from living in a strange, unfamiliar setting. From my experience, military families have a sense of peace about the threat of war or its reality. My biblical parallels drew attention to the worry, fear, grief, sorrow, anxiety, terror, and horror expressed across the social status throughout all ages by those found in similar situations. Are members of God's tactical unit exempt from emotional upset? Many of the Prophets, John the Baptist, the Apostles, and Paul and his companions would definitely offer a resounding "No!"

Historically, however, one of the most troublesome issues for families has been continual postings. Understandably, these would not be as great of a concern for unmarried members of the CF. This continual rolling-stone nature of military life does present a number of family difficulties, but when approached with an optimistic attitude they become opportunities. Many of these perceived difficulties are common to any move, whether civilian or military; but military relocations tend to be more frequent. Often a family leaves the amenities of one area, pulling up their roots every three or four years, and some may only put down very shallow, cedar-like roots to compensate. Arguments for and against the necessity of this gypsy-like life abound, but as pointed out, attitude plays a contributing factor in coping well. A pessimist or person with a 'what's-in-it-for-me' approach to life will find it an unnecessary trial on their family. This condition of employment makes it essential for couples contemplating marriage or cohabitation to begin the relationship with a firm foundation built on trust, communication, and lifestyle knowledge. When we considered marriage in a later chapter, these three proved indispensible to any marriage and lifestyle, but particularly important in this context. The intent of this chapter was to highlight existing parallels between families called to serve in God's tactical unit and those called to serve in the CF. Both must maintain a degree of portability.

The most stressful lifestyle issue is deployment, especially one sending a family member into a theatre of war. Deployments with varying degrees of home front stress include peacekeeping missions, international relief efforts, maneuvers and responses to natural disasters; these also prove our military is not just Canada's war machine. PTSD, called by a variety of names since the early 1600s, represents a potential challenge as troops return from any deployment. It is significant enough that it warranted further consideration in a separate chapter. We studied the seven stages of the three phases of deployment and discovered many home front and front line spouse frustrations. I concluded that long separations represent potential IEDs. Fortunately, human beings are capable of coping with periods of upheaval in their life and can repair foundational damage to rebuild new stability into a relationship over time. An understanding church can assist by removing as much stress from military families as possible during deployment.

"For better or for worse" is a phrase used in marriage vows; it was applied to depict the adult working years of life. I subdivided it into seven segments, with the first three being marriage-related. The primary point in all three is acceptance of people by the church just as they are, allowing leadership an opportunity to move them beyond that point. In this way, couples can grow spiritually and relationally while being taught and counseled. We learned that military social structure differs greatly from that of civilians, with Collier touching on numerous points. This was an interview valuable in assisting me to fathom lifestyle issues. Additional stress factors were considered, noting that the main stress may be a loved one deployed, but it is the daily aggravations that push emotions to the breaking point. Many of these issues could be alleviated by a church applying a literal rendering of the Sheep and the Goats Parable (Matthew 25:31–46). Culture was also included in this chapter, pointing out our military requires a type of evangelism similar to the approach used in youth ministry. There is also a need to be a risk-taker and to be creative in military ministry. In short, we can expect to experience an array of social challenges, cultural adjustments, and stress as we serve in God's tactical unit.

In chapter eight, the focus turned to children and senior citizens within the military community. Children require the most attention as one attempts to understand their feelings on family issues. This may be a difficult task but one well worthwhile to help them with family transitions. Senior citizens offer years of experience and deserve to live life with dignity and respect. Veterans are a segment of the senior population. I noted the treatment Veterans receive today reflects on us as a nation and on what our soldiers can expect in the future because Today's Trooper = Tomorrow's Veteran.

Remembrance Day opened my eyes to the value of supporting our veterans as a means of illustrating support for our troops. I also introduced a statement by a prominent evangelist, which bears repeating: "When you understand the people, you will often know what to say and do, and how."[174] By understanding people, we become equipped to show respect for veterans, children, and every individual we encounter.

In the chapter 'For Better or For Worse,' I implied that everyone experiences a little of both, but no one really looks forward to the 'For Worse' side of the equation. Certain facts about PTSD should be highlighted:

- "Post-traumatic Stress Disorder is a *normal* reaction to an *abnormal* event."[175]
- PTSD is not specific to any age or gender
- PTSD is a major international health concern
- PTSD is often talked about, but has not yet been aggressively acted upon by the Canadian government
- PTSD is more recognized, treated, and understood in the U.S.A.
- "Degree of Trauma $\times$ Social Support = Magnitude of Post-Traumatic Stress."[176]

---

[174] Hunter. 20
[175] Adsit. 23 (Author's emphasis)
[176] Grossman. 283

I believe Adsit's statement must always be kept in mind, as well as Grossman's equation. This is especially true in light of Floyd's comment: "A key factor that mitigates how a person responds to trauma is the individual's level of social support before, during, and following a traumatic occurrence. A person who has close, supportive relationships has a better likelihood of coping effectively."[177]

The most effective way for Christian civilian communities to assist in the PTSD battle is with supportive action. Individually and as a community, we have a responsibility to support our troops by becoming increasingly aware of the military in our midst and intensifying support for them.

Chaplain's Corner allowed three chaplain authors to share with us the ministry of chaplaincy in three areas. Before we took a detailed look at their work, I offered a brief overview of the beginning of chaplaincy and noted how modest individuals in this field of ministry tend to act. Our chaplains shared on burials, family communication, and serving as an unarmed soldier; some of this material was very descriptive in nature. As well, reading their comments carefully reveals other numerous duties performed. Chaplains carry a heavy burden as they live and work within military life. Because they are an invaluable support for the troops, it is our Christian responsibility to pray specifically for them and their families.

Before wrapping up this conclusion, I am reminded of a comment made in chapter one. In effect, I said if all your questions are not answered concerning my life, ministry, and military involvement, more information would surface throughout the balance of the chapter and book. I hope you found that to be the case. I also realize you may be pondering why God would call a military illiterate country boy to investigate and write about our CF and its need for ministry. It is an excellent question that I asked myself many times. There is no short, simple answer. What I discovered is a parallel between my call to the military and the call of another young country preacher from Pennsylvania. As David Wilkerson sought assistance from his congregation to reach out to New York street kids in 1958, he faced a similar question—the 'Why me?' question. He shares the call:

---

[177] Floyd. 60

I was dumbfounded by a thought that sprang suddenly into my head—full-blown, as though it had come into me from somewhere else.
Go to New York City and help those boys.
I laughed out loud. "Me? Go to New York City? A country preacher barge into a situation he knows less than nothing about?"
Go to New York City and help those boys. The thought was still there, vivid as ever, apparently completely independent of my own feelings and ideas.[178]

This call on Wilkerson's life led to the founding of Teen Challenge. His book depicts the interesting and challenging paths he travelled over the next few years. Each excursion into New York was a learning and occasionally scary trip—but one full of rewarding adventure for him. A parallel exists between our missions. My first call to assist in a military setting was a scary but learning occasion, in some ways similar to Wilkerson's first trip to New York City. I was being asked to barge into a situation I knew less than nothing about—our military. Allow me to paraphrase missionary Hudson Taylor's words regarding the Great Commission: Wilkerson and Darlington both responded to the Great Commission not as an option to be ignored but as a command to be obeyed. Love for God and neighbour propelled both of us forward.

I suspect two reasons prevail in our call to these distinct ministries. First, God seeks those who will be obedient to His leading. He knew Wilkerson would go to New York and I would go to the base. Second, I believe God seeks to accomplish His plans through ordinary people rather than know-it-all experts with a humanly devised ideal plan for ministry. Wilkerson admits he knew "less than nothing" about New York City and those boys when he was called. I soon discovered I knew "less than nothing" about our CF and its members prior to the fall 2007. As I reflect on God's call to both of us, His choice is sensible

---

[178] Reverend David Wilkerson. *The Cross And The Switchblade: The thrilling true story of a country preacher's fight against teen-age crime in big-city slums.* Old Tappan, NJ: Fleming H. Revell Company, 1963. 7 (Author's emphasis)

considering the target audience. In my opinion, both of us are targeting the average practical, hard-working, grassroots Christian or priest-in-the-pew. I surmise Wilkerson's audience and mine are likely as illiterate about our respective fields of ministry as we were when we began our journeys. Therefore, His plan was to use two individuals knowing "less than nothing" about these ministries and make them "somewhat informed" to allow them to communicate the needs and opportunities within these fields of ministry. God wanted two people to willingly become a somewhat informed 'priest-in-the-pew' and called them to share their ministry with the somewhat less informed 'priests-in-the-pew.' The term 'priest-in-the-pew' is in keeping with my understanding of the priesthood of all believers; I call ministers and pastors 'priests-in-the-pulpit.' Christians are both a member of God's tactical unit as well as a priest functioning under the direction of the Holy Spirit and the guidance of our High Priest Jesus Christ.

At the opening, I proposed to enrich the spirit of individuals in God's tactical unit. We are called to be *"a good soldier of Christ Jesus"* (2 Timothy 2:3). I wanted to raise awareness of the need for enhanced support for our troops within Christian civilian communities. These targets were attacked by aiming at existing parallels between two methods of warfare identified as spiritual and temporal. My strategic plan attacked the proposed objectives from two fronts. One front engaged paraphrased scriptural and non-scriptural poetic readings to activate the brain's rational side and trigger its creative side. Concurrently, a second invasion commenced awakening the spirit and soul to some identifiable parallels between biblical life and military life by engaging members of God's tactical unit from the Old and New Testaments. However, this has all been tainted by personal biases and emotions. This book is my personal account and reflects what my eyes, ears, and understanding have revealed to me about God, life, and the military. This has not been an objective but a subjective paralleling of two lifestyles; it is definitely not a scientific or sociological portrait.

Before a final biblical parallel, it is imperative to revisit my preliminary questions and raise them once again. Jesus made it clear only two options exist when He said, *"He who is not with me is against me, and he who does not gather with me scatters"* (Matthew 12:30). I began with these two questions: are you a Christian? If not, did you ever wonder what the Christian life might be like? I went on to provide the only three possible answers. Now I am asking you to again consider these two questions. For if this is a step you have yet to take, please consider it now for your soul's eternal sake.

Biblical Parallel

As we conclude, the closing comments of the book of Acts are applicable. Paul was in Rome with the freedom to preach:

> *"The Holy Spirit spoke the truth to your forefathers when he said through Isaiah the prophet:*
> *"'Go to this people and say, "You will be ever hearing but never understanding; you will be ever seeing but never perceiving." For this people's heart has become calloused; they hardly hear with their ears, and they have closed their eyes. Otherwise they might see with their eyes, hear with their ears, understand with their hearts and turn, and I would heal them.'*
> *"Therefore I want you to know that God's salvation has been sent to the Gentiles, and they will listen!"* (Acts 28:25–28)

Luke closes with this statement: *"For two whole years Paul stayed there in his own rented house and welcomed all who came to see him. Boldly and without hindrance he preached the kingdom of God and taught about the Lord Jesus Christ"* (28:30–31). This is how Luke chose to close the book he had written for Theophilus. Paul was still alive, preaching and teaching about the Lord Jesus Christ, but we have no idea what happened at the end of the two years Luke mentioned. Luke recorded the early history of the church from the birth of its founder through His death, burial, resurrection, ascension. and outpouring of the Holy Spirit and on through Paul's conversion and the establishment of local gatherings up until the late 60s A.D. Acts concludes in a very open-ended fashion, leaving someone else to complete the story. Why Luke ended his account at this point we do not know, but the continuation of church development is history. We may not know exact details of its growth, however, over two thousand years later the church remains waiting and watching for Christ's Second Coming (His Second Deployment). In parallel fashion, this book is closing in an open-ended fashion with the author still alive. I am preaching and teaching about the Lord Jesus Christ in a Pauline-style ministry, but have no idea what will happen from here. Just as Paul could never have imagined how God would use his faithful obedience, neither can I imagine how God will use my endeavours to serve Him. The same is true for you, dear reader. You cannot imagine how God will use your efforts on His behalf as part of His tactical unit. However, we can rest assured that our individual efforts do not go unnoticed or without reward in God's kingdom. In fact, there are no unknown soldiers in God's tactical unit. I believe sufficient evidence has been provided to challenge Christians to assess their position in God's tactical unit. Thus, I rest my case for military ministry and its ability to produce within us an enriched spirit as *"a good soldier of Christ Jesus"* (2 Timothy 2:3).

EPILOGUE

MY MANUSCRIPT IS COMPLETE, I AM ENJOYING VOLUNTEER MINISTRY IN retirement, and I have extra time for additional reading and research into military life. I've added my latest reading to the Bibliography before publication of the book, so the Bibliography is current. These additional sources have confirmed my belief that Christians require a primer for life as *"a good soldier of Christ Jesus"* (2 Timothy 2:3) and one designed to raise awareness within Christian civilian communities of the need for Canadians to enhance support for our troops. From this additional reading, one author's comments uniquely summarize and express my understanding of the work of numerous authors. Davis begins by pointing out that David Collenette, Minister of National Defence, called a news conference to announce, "Due to the systematic problems within the Airborne Regiment, it has been decided that the only possible solution to restore public confidence in our military is to disband the Canadian Airborne Regiment."[179]

James R. Davis began his military career in 1985 and had been "a member of the regiment for seven months"[180] when it was disbanded. On this action, he writes:

> I had an opportunity to speak with General Vernon...and asked the question..."General, where were the Public Affairs officers

---

[179] James R. Davis. *The Sharp End: A Canadian Soldier's Story.* Vancouver, BC: Douglas & McIntyre, 1997. 262
[180] J. Davis. 264

before the disbandment? Why weren't they out there flooding the media with good PR in our defence?"

The general looked at me and I could see I had struck a sore point.

"Sergeant, I have often asked myself the same question. The fact is, our Public Affairs branch was never created to put out good PR for anybody. It was designed from the start to do damage control for the senior brass," he replied.

I walked away from the general. In my mind that confirmed it. We had been set up and betrayed by our military masters. They had cooperated fully with the government's plan to get rid of the regiment, even before the minister had made his decision."[181]

In his concluding statements, Davis offers this insightful comment:

There must be an unwritten trust between a government and its soldiers. In return for offering to give up their lives to achieve the government's goals at home and abroad, the government must undertake to protect their soldiers and not throw their lives away needlessly. Looking out for their welfare must be a fundamental of building a reliable national army. If the soldiers lose faith in their leadership at a political level, disaster will follow.

…in disbanding the Airborne Regiment for political necessity, Canada's politicians broke faith with their soldiers.[182]

Regardless of one's opinion on disbanding of the Canadian Airborne Regiment, I believe Canada's politicians continue to break faith with our soldiers. Recently, the government has failed to protect our troops in at least two areas: 1) implementing lump-sum payments for injuries and 2) their sluggish response to PTSD treatment and support. Thus, the Christian church in Canada has a moral obligation to seek justice for

---

[181] J. Davis. 263-4
[182] J. Davis. 269

those "offering to give up their lives to achieve the government's goals at home and abroad." The starting point for visibly demonstrating support for our troops is through an understanding of military life. Once we understand their culture, troop support will naturally flow from thankful hearts. A list of various support organizations follows for the reader to consider:

http://www.boomerslegacy.ca
http://www.equitassociety.ca
http://wwwifishouldfall.com
http://www.projecthealingwaters.ca
http://teamredtakeastand.com
http://www.vetscanada.org
http://www.woundedwarriors.ca

APPENDICES

Appendix A
Local Christian Civilian Community Survey

Research Questionnaire

It is my hope this questionnaire will be completed by a member of the ministerial staff or a church leader (i.e. Chair of Elders, Mission Committee, etc.). Also please note:
1) all feedback is welcome and can only serve to enhance this investigative work;
2) feel free to remain anonymous, although church identity would be appreciated;
3) you are only asked to complete the sections which apply to your ministry situation.

Section A – Identification
Respondent's Name:_____
Position & Church: _____
Contact Information: _____
_____
_____

Section B – Your Thoughts on Military Ministry
Please indicate your level of agreement with the following statements.

a) Our Military constitutes a special and unique mission field.
   Disagree   1.   2.   3.   4.   5.   Agree

b) The universal Christian church in Canada should be more supportive of ministry to our troops.
   Disagree   1.   2.   3.   4.   5.   Agree

c) Congregations with an interest in ministering to our military should have a specific plan or program to reach out and assist military personnel and their families.
   Disagree   1.   2.   3.   4.   5.   Agree

d) Our military should be served through a specially designated military ministry just as many congregations provide a special youth ministry.
   Disagree   1.   2.   3.   4.   5.   Agree

e) It would be beneficial for both our congregation and our troops if more information was available on ways our congregation could support our troops.
   Disagree   1.   2.   3.   4.   5.   Agree

f) Our congregation would be receptive to a seminar focused on the needs and lifestyle of military personnel and their families.
   Disagree   1.   2.   3.   4.   5.   Agree

g) Our congregation would be interested in developing or expanding its ministry to our military.
   Disagree   1.   2.   3.   4.   5.   Agree

h) Ministry to military personnel and their families would benefit from support and promotion by military personnel and their families presently attending our church.
    Disagree 1. 2. 3. 4. 5. Agree

i) The church is "…simply not sensitized to the unique needs of military personnel, or they are somewhat intimidated by them, particularly by returning warriors…" (Robert F. Dees, Major General, U.S. Army, Retired)
    Disagree 1. 2. 3. 4. 5. Agree

j) It would be beneficial to congregations if a coordinator was available to act in a communication role between the local Christian civilian community and Base officials.
    Disagree 1. 2. 3. 4. 5. Agree

Section C – Sharing
I) What You are Doing:
If your congregation currently has ministry directed toward military personnel and their families, either exclusively or jointly with civilians, please describe the past six months or so of this ministry, listing ongoing programs and events with an assessment of them.

II) Your Strengths:
If your congregation were to invest in an intentional ministry to our military, or if it were to expand a current ministry, please address the following questions:
1. Can you identify and list new ministry initiatives you believe your congregation could develop to support our troops? (By this I refer to "acts of Christian service" or visible, tangible assistance).

2. Although we may desire to reach the lost, not every congregation is gifted for all it may desire to accomplish. Can you identify and list areas of strength in your congregation which would

help you in developing an "acts of Christian service" ministry in support of our military?

III) What's Working – What's Not?

Reflecting on past events in your congregation directed at demonstrating support for our troops, were they well attended? Please comment on why you believe they were or were not well attended including whether or not you feel promotion on the Base may have been beneficial.

Section D – Additional Thoughts and Comments
o   Please check if you are interested in receiving a copy of the completed study.

Appendix B
Acronyms Used In Military Language

AAV (or Tracks) – Amphibious Assault Vehicle
AMCF – Association of Military Christian Fellowships
ANA – Afghan National Army
AWOL – Absent without Leave
CBC – Canadian Broadcasting Corporation
CF – Canadian Forces
CFB/CFBs – Canadian Forces Base/Canadian Forces Bases
CO – Commanding Officer
Combats – Green and brown camouflage uniforms that are frequently seen
CP/CPs – Chaplaincy Program/Chaplaincy Programs
DND – Department of National Defence
HQ – Headquarters
Humvees (HMMWVs) – High Mobility Multipurpose Wheeled Vehicles
IED/IEDs – Improvised Explosive Device/Improvised Explosive Devices
ICCMC – Interfaith Committee of Canadian Military Chaplaincy
KIA – Killed-in-Action
LAV/LAVs – Light Armoured Vehicle/Light Armoured Vehicles
LCol – Lieutenant-Colonel (Canadian)
Lt.Col – Lieutenant-Colonel (American)
MA – Medical Assistant
MGen. – Major-General
MCF – Military Christian Fellowship
MCF-Pet – Military Christian Fellowship - CFB Petawawa chapter
MCF Rep-Pet – Individual representing MCF-Pet. at the MCF national level
MFRCs – Military Family Resource Centres
MO – Medical Officer
NARC – North American Regional Conference
NCO – Non-Commissioned Officer

NOK – Next of Kin
PACA – Pembroke and Area Clergy Association
PEN – Personal Emergency Notification
PMQs – Permanent Married Quarters
PT gear – Physical Training gear; other options combats or full dress uniform
PTSD – Post-Traumatic Stress Disorder
STS – Secondary-Traumatic Stress

Appendix C
Definition of Terms

Terms applied in this book are defined as follows:

Base or Canadian Forces Base: the home region of various groupings of military personnel and their families. There are numerous CFBs across our nation.

Christian civilian community: a gathering of Jesus Christ's disciples—*"The disciples were called Christians first at Antioch"* (Acts 11:26). Civilian means the Christian gatherings consist primarily of non-military individuals and their families; however, it does not exclude military individuals and their families who may be part of a local gathering. Community refers to gatherings, not geographic location. A community is the gathering of individuals and families for teaching and promoting the Gospel of Jesus Christ. Thus, community locations include but are not limited to: church buildings, Christian education centres, and privately owned facilities.

Civilian soldier: the collective occupations of law enforcement officers, fire fighters, emergency response personnel, security guards, prison staff, and other professionals with an above-average risk of death or physical injury.

Conscription: compulsory military service required by the national government from its citizens. There is no choice. It stands in contrast to enlisting which is voluntary commencement of military training.

Deployment: the assignment of military personnel for duty, unaccompanied by their family. This duty or engagement may include an off-base course, an off-base training exercise, a mission overseas, or serving in the event of a national disaster either at home or abroad. For an assignment to be classified as a deployment, military personnel must be separated from their loved ones for thirty days or more.

Element of the Armed Forces: a term referring to what many will remember as a branch of the military. The three branches or elements of our military were once listed as Air Force (air), Army (ground), and Navy (water).

Enlist: voluntarily signing up to commence military training. To enlist is a choice one makes to join the military; there is no obligation to make such a choice. Conscription, by contrast, is mandatory military service required by the government.

Esprit de corps: loyalty and devotion uniting the members of a group (French)[183]

Leave: an extended period of time off-duty for military personnel, similar to civilian vacation or holidays. Generally there is little choice in selecting the time one can take as leave. It is scheduled around military demands, not personal desires or wishes.

Military family: a military family is any individual or group of individuals where at least one person is enlisted in our Canadian Forces (CF). Thus, a single soldier is a family unit in the broadest sense. Furthermore, in most cases civilians employed on the base are not considered part of a military family. They are employed personnel, not enlisted.

Non-commissioned: a military term for positions not at the commissioned or senior-officer level.

Operational Stress Injury (OSI): "Any persistent psychological difficulty resulting from operational duties performed in the course of military service. OSI is a more comprehensive term than PTSD; it may be thought of as an umbrella term for PTSD, other anxiety disorders and depression. It re-characterizes these conditions as injury, which is more in keeping with current thinking. OSI is not a legal or a medical

---

[183] Hawkins. 272

term. Unlike PTSD, it is a strictly military term, used by Canada and NATO."[184]

Pembroke and Petawawa Christian civilian community: a reference to the communities of Pembroke, Petawawa, Chalk River, Deep River and area.

Posting: a military term used to describe the relocation of military personnel from one base to another (for example a family being transferred from CFB Petawawa to CFB Borden is said to have been posted to Borden or received a posting to Borden).

Post-Traumatic Stress Disorder (PTSD): "An anxiety disorder caused by an experience in which serious physical harm or death occurred or was threatened."[185]

Priest-in-the-pew: a personal term of reference for individual believers who sit in church pews and make up a congregation. I view the pastor as the priest-in-the-pulpit; the terms are applied to endorse my belief in the concept of the priesthood of all believers.

Support: when used as a reference in ministry to our troops, support reflects the combination of two verbs—"to learn" and "to serve."

Temporal (*adj.*) 1. secular, of worldly affairs as opposed to spiritual[186]

Tour: another word applied to a deployment in the sense of a tour of duty.

---

[184] http://www.forces.gc.ca/site/news-nouvelles/view-news-afficher-nouvelles-eng.asp?id=2871
[185] http://www.forces.gc.ca/site/news-nouvelles/view-news-afficher-nouvelles-eng.asp?id=2844
[186] Hawkins. 842

BIBLIOGRAPHY

Books

Adsit, Rev. Chris. *Combat Trauma Healing Manual: Christ-centered Solutions for Combat Trauma*. Newport News, VA: Military Ministry Press, 2007.

Adsit, Chris, Rahnella Adsit and Marshele Carter Waddell. *When War Comes Home: Christ-centered Healing for Wives of Combat Veterans*. Newport News, VA: Military Ministry Press, 2008.

Almond, Johnny, Rob Edwards and Kermit Jones, Jr. comp. *Ministry To The Military In Our Midst*. Printable document available at http://ministeringtomilitary.org

Airth, Lesley Anne. *What We Remember*. Renfrew, ON: General Store Publishing House, 2004.

Armstrong, Keith, Suzanne Best and Paula Domenici. *Courage After Fire: Coping Strategies for Troops Returning from Iraq and Afghanistan and Their Families*. Berkeley, CA: Ulysses Press, 2006.

Barner, Stefani E. *Faith and Magick in the Armed Forces: A Handbook for Pagans in the Military*. Woodbury, MN: Llewellyn Publications, 2008.

Bercuson, David. *Significant Incident: Canada's Army, the Airborne, and the Murder of Somalia*. Toronto, ON: McClelland and Stewart Inc, 1996.

Boyle, Everett. *The Rest of the Story: According to Boyle*. Burnstown, ON: General Store Publishing House, 2002.

Carroll, Andrew, ed. *Operation Homecoming: Iraq, Afghanistan, and the Home Front, in the words of U.S. Troops and Their Families*. New York: Random House, 2006.

Cash, Lt. Carey H. *A Table in the Presence: The Dynamic Account of How a U.S. Marine Battalion Experienced God's Presence amidst the Chaos of the War in Iraq*. Nashville: Thomas Nelson, 2004.

Cole, Darrell. *When God Says War Is Right: The Christian's Perspective on When and How to Fight*. Colorado Springs, CO: WaterBrook Press, 2002.

Coleman, Lyman R. *In this Sign: An Historical Novel*. Renfrew, ON: General Store Publishing House, 2005.

Collier, Dianne. *Hurry Up and Wait: An Inside Look at Life as a Canadian Military Wife*. Carp, ON: Creative Bound Inc, 1994.

_____. *My Love, My Life: An inside look at the lives of those who love and support our military men and women*. Carp, ON: Creative Bound Inc, 2004.

Colombo, John Robert and Michael Richardson, compilers. *We Stand On Guard: Poems and Songs of Canadians in Battle*. Toronto, ON: Doubleday Canada Limited, 1985.

Cook, Ramsay, Craig Brown and Carl Berger, ed. *Conscription 1917: Canadian Historical Readings*. Essay by J.M. Bliss. "The Methodist Church and World War I". Toronto, ON: University of Toronto Press, nd.

Curry, Dayna and Heather Mercer. *Prisoners Of Hope: The Story of Our Captivity and Freedom in Afghanistan*. Colorado Springs: WaterBrook Press, 2002.

Doucette, Fred. *Empty Casing: A Soldier's Memoir of Sarajevo Under Fire*. Toronto, ON: D & M Publishing Inc, 2008.

Davis, Eldon S. *An Awesome Silence: A Gunner Padre's Journey Through The Valley Of The Shadow*. Carp, ON: Creative Bound Inc, 1991.

Davis, James R. *The Sharp End: A Canadian Soldier's Story*. Vancouver, BC: Douglas & McIntyre Ltd, 1998.

Ellis, Deborah. *Off to War: Voices of Soldiers' Children*. Toronto, ON: Groundwood Books, 2008.

Fishback, Beatrice. *Loving Your Military Man: A Study for Women Based on Philippians 4:8*. Little Rock, AR: FamilyLife Publishing, 2007.

Fishback, Lt. Col. Jim, USA (Ret.) and Beatrice Fishback. *Defending the Military Family*. Little Rock, AR: FamilyLife Publishing, 2005.

_____. *Defending the Military Marriage*. Little Rock, AR: FamilyLife Publishing, 2005

Floyd, Scott. *Crisis Counseling: A Guide For Pastors and Professionals*. Grand Rapids: Kregel Publications, 2008.

Fung, Mellissa A. *Under An Afghan Sky: A Memoir of Captivity*. Toronto, ON: HaperCollins Publisher Ltd, 2011.

Gordon, Earnest. *To End All Wars: A True Story about the Will to Survive and the Courage to Forgive*. Grand Rapids: Zondervan, 1963.

Grant, John Webster. *The Church in the Canadian Era*. Burlington, ON: Welch Publishing Company Inc, 1988.

Green, Jocelyn. *Faith Deployed: Daily Encouragement for Military Wives*. Chicago: Moody Publishers, 2009.

Greene, Bob. *Once Upon A Town: The Miracle of North Platte Canteen*. New York: HarperCollins Publisher, 2002.

Grossman, Lt. Col. Dave. *On Killing: The Psychological Cost of Learning to Kill in War and Society*. New York: Black Bay Books, 1996.

Hadley, Donald W. and Gerald T. Richards. *Ministry with the Military: A Guide for Churches and Chaplains*. Grand Rapids: Baker Book House, 1992.

Harrison, Deborah, and Lucie Laliberté. *No Life Like It: Military Wives in Canada*. Toronto, ON: James Lorimer & Company, 1994.

Hays, Richard B. *Echoes of Scripture in the Letters of Paul*. New Haven, CT: Yale University Press, 1989.

Henderson, Kristin. *While They're at War: The True Story of American Families on The Homefront*. New York, NY: Houghton Mifflin Company, 2006.

Holliday, Alesia. *E-mail to the Front: One Wife's Correspondence with Her Husband Overseas*. Kansas City, MO: Andrews McMeel Publishing, 2003.

Hope, LCol. Ian. *Dancing with the Dushman: Command Imperatives for the Counter-Insurgency Fight in Afghanistan*. Kingston, ON: Canadian Defence Academy Press, 2008.

Horn, LCol. Bernd. *Bastard Sons: An Examination Of Canada's Airborne Experience 1942–1995*. St. Catharines, ON: Vanwell Publishing Ltd, 2001.

Hunter, George G. III. *The Celtic Way Of Evangelism: How Christianity Can Reach The West...Again*. Nashville: Abingdon Press, 2000.

Kallin, Anne. *Proudly She Marched: Training Canada's World War II Women In Waterloo County*. Volume 2: Women's Royal Canadian Naval Service. Waterloo, ON: Canadian Federation of University Women, 2007.

Kay, Ellie. *Heroes At Home: Help & Hope for America's Military Families*. Grand Rapids: Bethany House, 2008.

Kubler-Ross, Elisabeth. *On Death and Dying*. New York: Macmillan Publishing, 1969.

Justice, Cpl. Nathan. *Path of the Warrior: Spiritual Lessons from the Front Line*. Winnipeg, MB: Word Alive Press, 2010.

Lucado, Max. *Facing Your Giants: A David and Goliath Story for Everyday People*. Nashville: Thomas Nelson, 2006.

MacKenizie, MGen. Lewis. *Peacekeeper: The Road to Sarajevo*. Vancouver, BC: Douglas & McIntyre, 1993.

Marchand, Christopher. *Restoring Rebecca: A story of traumatic stress, caregiving, and the unmasking of a superhero*. Denver, CO: Outskirts Press Inc, 2009.

Monetti, Tony Lt. Col. and Penny Monetti. *Called to Serve: Encouragement, Support, and Inspiration for Military Families*. Grand Rapids, MI: Discovery House Publishers, 2011.

Montgomery, Mike and Linda, and Keith and Sharon Morgan. *Making Your Marriage Deployment Ready*. Little Rock, AR: FamilyLife Publishing, 2008.

Murphy, Terrence. ed. *A Concise History of Christianity in Canada*. Don Mills, ON: Oxford University Press, 1996.

Neven, Tom. *On the Frontline: A Personal Guidebook for the Physical, Emotional, and Spiritual Challenges of Military Life*. Colorado Springs, CO: WaterBrook Press, 2006

Ogle, James and Darnell Bass. *What Manner of Man: Darnell Bass and the Canadian Airborne Regiment*. Renfrew, ON: General Store Publishing House, 2006.

Pinch, Franklin C, et al. edited by. *Challenge and Change in the Military: Gender And Diversity Issues*. Kingston, ON: Canadian Defence Academy Press, 2006.

Pavlicin, Karen M. *Surviving Deployment: A guide for military families*. Saint Paul, MN: Elva Resa Publishing, 2003.

Rawlyk, George A. ed. *The Canadian Protestant Experience 1760–1990*. Kingston, ON: McGill-Queen's University Press, 1990.

Russell, Jeanette (Shetler). *We're Moving Where? The life of a Military Wife*. Renfrew, ON: General Store Publishing House, 2003.

Ruth, Peggy Joyce. *Psalm 91: God's Shield of Protection*. No publisher information provided, contact: www.peggyjoyceruth.org, 2007.

Sherman, Michelle D. and DeAnne M. Sherman. *Finding My Way: A Teen's Guide to Living with a Parent Who Has Experienced Trauma.* Edina, MN: Beaver's Pond Press, Inc, 2005.

_____. *My Story: Blogs by Four Military Teens.* Edina, MN: Beaver's Pond Press, Inc, 2009.

Snailham, Jane. *Eyewitnesses to Peace: Letters from Canadian Peacekeepers.* Clementsport, NS: The Canadian Peacekeeping Press, 1998.

Starr, Judy. *The Enticement of the Forbidden: Protecting Your Marriage.* Peachtree City, GA: LifeConneXions, 2004.

_____. *The Enticement of the Forbidden: Personal Study and Discussion Guide.* Peachtree City, GA: LifeConneXions, 2004.

Summerby, Janice. *Native Soldiers—Foreign Battlefields.* Remembrance Series, Publication of Ministry of Veteran Affairs, Canada, 2005.

Taylor, Dianne J. *There's No Wife Like It.* Victoria, BC: Braemar Books Ltd, 1985.

Vandesteeg, Carol. *When Duty Calls: A Handbook for Families Facing Military Separation.* Colorado Springs, CO: David C. Cook, 2005.

Waddell, Marshele Carter. *Hope for the Home Front: Winning the Emotional and Spiritual Battles of a Military Wife.* Birmingham, AL: New Hope Publishers, 2003.

_____. *Hope for the Home Front: Winning the Emotional and Spiritual Battles of a Military Wife, Bible Study.* Birmingham, AL: New Hope Publishers, 2005.

Walters, Eric. *Wounded*. Toronto, ON: Penguin Group (Canada), 2009. (A novel)

Watkins, Morris G. and Lois I. Watkins, Sen. Ed. *The Complete Christian Dictionary For Home and School*. Ventura, CA: Gospel Light, 1992.

Wilkerson, David. *The Cross and the Switchblade: The thrilling story of a country preacher's fight against teen-age crime in big-city slumps*. Old Tappan, NJ: Fleming H. Revell Company, 1963.

Wright, H. Norman. *Crisis & Trauma Counseling: A Practical Guide for Ministers, Counselors and Lay Counselors*. Ventura, CA: Regal Books, 2003.

Other Media Resources

Brochures

Directorate of Medical Policy 2000. *Preparing For Deployment Stress*. Ottawa, ON: National Defence Publication, 2000.

_____. *Preparing For Reunion Stress*. Ottawa, ON: National Defence Publication, 2000.

Military Ministry Publication. *Church Guide for Ministering to the Military: An Introduction to the Bridges of Healing Ministry Including How to Provide Spiritual Care for Combat Trauma*. Newport News, VA: Military Ministry, n.d.

Ministry of Veteran Affairs. *Aboriginal Canadians In The Second World War*. Canada Remembers, n.d.

Ministry of Veteran Affairs, *Aboriginal Veterans*. Canada Remembers, 2005.

Petawawa Military Family Resource Centre Publication: Deployment Support Programs

Petawawa Military Family Resource Centre Publication: Preparing For Deployment Stress

Petawawa Military Family Resource Centre Publication: Reservist Soldiers!

Videos

Military Wives. A CTV Documentary by Triad Film Productions, 1999. Executive Producer: Peter d'Entremont. The video was inspired by the Harrison and Laliberté book *No Life Like It: Military Wives in Canada*. Toronto, ON: James Lorimer & Company, 1994. (Borrowed from D. Collier)

Military Wives – Celebrating the Life. Copyright: Debert Military Family Resource Centre, 2000. Produced by Video Tech, Halifax, N.S. (Borrowed from D. Collier)

Military Ministry, American Association of Christian Counselors & Light University. *Care & Counsel for Combat Trauma: Certificate Training Program*. Training Manual, 30 One-Hour DVDs, and Examinations for each of the 5 Units. Newport News, VA: Military Ministry Press, 2009. (Purchased)

Websites of Interest

Jones, Jennifer. "Tim Hortons in Kandahar, Afghanistan: An insider's view." http://www.canadianliving.com/life/community/tim_hortons_in_kandahar_afghanistan_an_insiders_view.php; accessed 20 March 2013.

http://www.cbc.ca/canada/story/2008/11/07/f-remembrance-day.html

http://www.cbc.ca/health/story/2008/12/17/military-stress.html

http://www.cmfamilies.ca

http://www.cmp-cpm.forces/gc.ca/cen/pub/cam-map/cha-01-eng.asp

http://www.cmp-cpm.forces/gc.ca/cfcb-bsafc/pub/iccmch-gciamc-eng.asp

http://commissionaires.ca

http://www.forces.gc.ca/site/news-nouvelles/view-news-afficher-nouvelles-eng.asp?id-2831

http://www.forces.gc.ca/site/news-nouvelles/view-news-afficher-nouvelles-eng.asp?id-2844

http://www.forces.gc.ca/site/news-nouvelles/view-news-afficher-nouvelles-eng.asp?id-2871

http://www.forces.gc.ca/site/news-nouvelles/view-news-afficher-nouvelles-eng.asp?id-3015

http://www.militaryminds.ca
www.facebook.com/militarymind-syt

http://www.ministeringtomilitary.org

http://www.osiss.c

http://www.peggyjoyceruth.org